Tropical
Rain Forests
and the
World Atmosphere

AAAS Selected Symposia Series

Published by Westview Press, Inc.
5500 Central Avenue, Boulder, Colorado

for the

American Association for the Advancement of Science
1333 H Street, N.W., Washington, D.C.

Tropical Rain Forests and the World Atmosphere

Edited by Ghillean T. Prance

AAAS Selected Symposium **101**

AAAS Selected Symposia Series

This book is based on a symposium that was held at the 1984 AAAS Annual Meeting in New York City, New York, on May 27. The symposium was sponsored by the New York Advisory Committee.

This Westview softcover edition was manufactured on our own premises using equipment and methods that allow us to keep even specialized books in stock. It is printed on acid-free paper and bound in softcovers that carry the highest rating of the National Association of State Textbook Administrators, in consultation with the Book Manufacturers' Institute.

Published in 1986 in the United States of America by Westview Press, Inc.; Frederick A. Praeger, Publisher; 5500 Central Avenue, Boulder, Colorado 80301

Library of Congress Cataloging-in-Publication Data
Prance, Ghillean T., 1937-
    Tropical rain forests and the world atmosphere.
    (AAAS selected symposium ; 101)
    1. Forest meteorology. 2. Rain forest ecology.
3. Deforestation--Environmental aspects.
4. Deforestation--Tropics. 5. Atmosphere.
6. Vegetation and climate. 7. Forest meteorology--
Tropics. 8. Vegetation and climate--Tropics.
I. Title. II. Series.
SD390.5.P73 1986   551.6   85-31521
ISBN 0-8133-7176-7

Composition for this book was provided by the editor.
This book was produced without formal editing by the publisher.

Printed and bound in the United States of America

The paper used in this publication meets the requirements of the American National Standard for Permanence of Paper for Printed Library Materials Z39.48-1984.

6   5   4   3   2

About the Book

Public awareness of the advancing destruction of tropical forest habitats has grown in recent years, as has the scientific understanding of the complexity of these diverse and fragile ecosystems and the importance of their contribution to the global atmosphere. Current research demonstrates that the role of tropical forests in maintaining the equilibrium of the atmosphere may be far greater than previously believed and that the accelerating rate of forest destruction may have profound implications for atmospheric budgets of N_2O, CH_4, CO_2, and important trace gases. Large-scale deforestation may also have serious and unforeseen effects on climate and hydrology.

New techniques such as remote sensing have made possible accurate evaluations of the rate of destruction of tropical rain forests. The results are alarming: Norman Myers, in Chapter 2, estimates that 200,000 square kilometers of this biome are being depleted each year from a total area of less than 10 million square kilometers.

This volume, the result of a AAAS symposium organized to explore the ramifications of tropical deforestation, emphasizes the relationship of biosphere to troposphere, aiming to set tropical forest ecology in the context of the global ecosystem. Case studies illustrate our increasing knowledge of these important habitats and the urgency of finding ways to preserve them. Diagnoses are accompanied by prescriptions for future policies.

About the Series

The *AAAS Selected Symposia Series* was begun in 1977 to
provide a means for more permanently recording and more
widely disseminating some of the valuable material that is
discussed at the AAAS Annual National Meetings. The volumes
in this series are based on symposia held at the Meetings
that address topics of current and continuing significance,
both within and among the sciences, and in the areas in which
science and technology have an impact on public policy. The
series format is designed to provide for rapid dissemination
of information, so the papers are reproduced directly from
camera-ready copy. The papers are organized and edited by the
the symposium arrangers who then become the editors of the
various volumes. Most papers published in the series are
original contributions that have not been previously
published, although in some cases additional papers from
other sources have been added by an editor to provide a more
comprehensive view of a particular topic. Symposia may be
reports of new research or reviews of established work,
particularly work of an interdisciplinary nature, since the
AAAS Annual Meetings typically embrace the full range of
the sciences and their societal implications.

WILLIAM D. CAREY
Executive Officer
American Association for
the Advancement of Science

Contents

Tables

Figures

* Photo section follows page 38.

About the Editor and Authors

Ghillean T. Prance is senior vice president and director of the Institute of Economic Botany of the New York Botanical Garden. Since 1964 he has traveled extensively throughout the Amazon region carrying out a botanical survey of Brazilian Amazonia. He is author of over 150 scientific articles and monographs, mainly about Amazonian botany. He is editor of Extinction is Forever (New York Botanical Garden, 1977), Biological Diversification in the Tropics (Columbia University Press, 1982), and co-editor of Key Environments: Amazonia (with Thomas E. Lovejoy; Pergamon Press, 1985).

Richard A. Houghton is assistant scientist at the Ecosystems Center of the Marine Biological Laboratory at Woods Hole, Massachusetts. He has written scientific articles on the history of deforestation, the use of satellites to measure rates of deforestation, metabolism of terrestrial ecosystems, the global carbon cycle, and the effect of the biota on the carbon dioxide concentration.

Mark Leighton is assistant professor in the Department of Anthropology at Harvard University. His specialty is tropical ecology, and he has spent four years in Indonesia studying the evolutionary and ecological interactions between seeds and fruits of tropical rain forest plants and the vertebrates that feed on them.

Thomas E. Lovejoy, an ornithologist with extensive Amazon experience, is executive vice president of the World Wildlife Fund (U.S.), and director of the Critical Size Ecosystem Research Project of Amazonia sponsored jointly by the Instituto Nacional de Pesquisas da Amazonia (INPA) and the World Wildlife Fund. He has published extensively on Amazonian ecology.

Michael B. McElroy is Abbott Lawrence Rotch Professor of Atmospheric Sciences in the Division of Applied Sciences at Harvard University. His many honors and awards include a Public Service Award from NASA, the McIlwane Award of the American Geophysics Union, and the Newcomb Cleveland Prize of the American Association for the Advancement of Science. He is the author of more than 150 publications ranging from studies of other planets to issues affecting our global environment.

Norman Myers is an environmental consultant who specializes in tropical forests. He has consulted for many governments and international organizations and is the recipient of a Gold Medal from the World Wildlife Fund International. His books include The Sinking Ark (Pergamon Press, 1979), Conversion of Tropical Moist Forests (National Research Council, 1980), and The Primary Source: Tropical Forests and Our Future (Norton, 1984).

Gerard Piel, chairman of the board of Scientific American, has a distinguished career in science journalism. He is president (1985-1986) of the American Association for the Advancement of Science, a member of the board of managers of the New York Botanical Garden, and a member of many other scientific organizations, including the American Academy of Arts and Sciences. He is the author of Science in the Cause of Man (A.A. Knopf, 1962) and The Acceleration of History (A. A. Knopf, 1972).

Eneas Salati, former director of the Institute Nacional de Pesquisas da Amazonia (INPA), is currently director of the Centro de Energia Nuclear na Agricultura (CENA) at the University of São Paulo, Brazil. He is the author of numerous papers on the effect of deforestation on Amazonian rainfall.

Thomas A. Stone is research associate at the Ecosystems Center of the Marine Biological Laboratory at Woods Hole, Massachusetts. He specializes in remote sensing and he has written articles on measurement of deforestation by remote sensing, LANDSAT mapping of copper mineralization in Northeastern Brazil, and remote sensing of geobotanical anomalies.

Peter B. Vose is an adviser and project manager for the Centro de Energia Nuclear na Agricultura (CENA) at the University of São Paulo, Brazil. His principal area of research has been genetical effects in plant nutrition and also nuclear techniques in plant and soil sciences. He is the author of more than eighty scientific papers.

Nengah Wirawan is a botanist from Haganuddin University, Ujung Pandang, Indonesia. He has done extensive research on the effects of the recent fires that decimated a large area of the rain forests of Borneo.

Steven C. Wofsy is a senior research fellow in the Division of Applied Science at Harvard University. A specialist in atmospheric chemistry, he has spent several years studying trace gases in the Amazon region of Brazil.

George M. Woodwell is director of the Woods Hole Research Center in Woods Hole, Massachusetts. His numerous awards and honors include the Green World Award of the New York Botanical Garden and a Distinguished Service Award from the American Institute of Biological Sciences. He is the author of more than 200 publications including numerous papers on the carbon balance of the world's atmosphere.

Foreword

Few places on earth seem more distant from the temperate-zone city-dweller than the tropical rain forests. Yet events transpiring in that remote biome are fixing, perhaps unalterably, the prospects for mankind.

As we ought to know from the composition of our own bodies -- we are more air than dust -- atmospheric gases and vapors compose 90% of the substance of the biosphere, by weight and volume, taking up in living tissue no more than pinches of other elements from the lithosphere. In the exchange of gases and vapors between the interpenetrating biosphere and atmosphere, tropical rain forests play a role much larger than their acreage in ratio to the whole. As Michael B. McElroy and Steven C. Wofsy show in their contribution to this volume, fully half of the nitrous oxide, nearly half of the carbon monoxide, and a quarter of the methane released into the atmosphere emanate from the metabolism of tropical rain forests. To that metabolism human activity is contributing in mounting proportion.

More than half of the flux of carbon monoxide from tropical rain forests comes from the clearing and burning that is reducing their acreage all around the world. Carbon monoxide, methane, and nitrous oxide are of interest because they are catalysts of the reactions in the upper atmosphere that are tending to thin out the life-shielding ozone layer. The burning and decay of the tropical biomass now injects also as much carbon dioxide into the atmosphere as the burning of fossil fuels in the industrial nations of the Northern Hemisphere. By opaquing the infra-red window in the sky, the increasing concentration of this gas is blocking the re-radiation of the sun's heat from the earth. The destabilizing of the climate worldwide threatened by this development offers the ultimate argument for the necessity of nuclear power, put forward so compellingly elsewhere by George M. Woodwell (1).

Thus the outlook for temperate-zone city-dwellers and for their grandchildren is inextricably linked to the disappearance of the rain forests. The impoverishment of the life in the mid-latitudes and the extinction of plant and animal communities not yet named and catalogued might otherwise be regarded as a local event to be regretted outside only by taxonomists. Ghillean T. Prance develops, however, the poignance of this event. For the rain forests and for systematic biology he argues furthermore that the world economy will someday have to regret the loss of many plants of use to the indigenous peoples and of unrealized economic promise for a growing world population.

Eneas Salati and his colleagues exhibit another consequence of the disappearance of the rain forests that may reach into the world outside. In the upper Amazon the rain forest recycles the rains brought onto the continent from the Atlantic Ocean by the easterly trade winds. For the region as a whole, transpiration and evaporation supply fully half the rainfall; in basins with high forest cover, all but 20% of the local rainfall. The drought attendant on the disappearance of the Amazon rain forest -- 30% of the land area of the 20-degree equatorial belt -- can change the energy balance governing the planetary atmospheric circulation system. Associated with the deforestation of the Congo and the saharization-sahelization of equatorial Africa, this prospect urgently invites the cooperative international study necessary to raise the priority for preventative action before it is too late.

In relation to these ongoing and impending events, the fortunate denizens of the Northern Hemisphere are not only innocent bystanders; they are accomplices as well. Norman Myers shows that the rain forest in Latin America is losing 20,000 hectares a year to the extension of pasture for beef cattle. This is a consequence of the vertical integration of the marketing of hamburgers by fast-food chains in the United States that has been reducing traffic in the country's supermarkets. The persisting colonial relationship between the North and South has larger consequences. All around the 20-degree equatorial belt, from Africa, across the Amazon valley, through the

Archipelago of the Southwest Pacific into Africa again, "forest farmers" are clearing nearly 200,000 hectares of rain forest every year. This calamity is visible from outer space, as George M. Woodwell shows, to the cameras of the LANDSAT satellites that give it quantitative measurement. Destruction of the rain forests is the price being paid for the human population growth that is sustained by subsistence agriculture in its historic equilibrium with misery. Because the fragile soils -- in the Amazon, last glaciated 300 million years ago -- are themselves sustained by their plant and animal communities, they speedily go sterile when forest is cleared. The farmer thereupon takes his primitive technology deeper into the forest.

Four decades after the end of World War II, the promise of Freedom From Want is put down as an earlier generation's hypocrisy. Only two nations, Canada and Sweden, have met the undertaking subscribed to by all the industrial nations in two unanimous votes of the United Nations General Assembly to commit 1% and then 0.7% of their gross national product to accelerate the industrialization of the pre-industrial underdeveloped nations.

Human want now comes to the fore as a planetary environmental force. Over thousands of millennia, the rain forests have survived oscillations of the world climate more extreme than any experienced in historic time. As the anthropologist Mark Leighton shows, however, the rain forest of the Southwest Pacific archipelago may not survive the present cycle of drought brought on by the recurrent die-away of the Southeast Pacific northward-running Antarctic-Peruvian current, known as the El Niño Southern Oscillation. The 1982-83 El Niño had severe ecological consequences throughout the Pacific basin. The drought-stricken Borneo forest is being swept by wild fire, ignited on all sides by the slash and burn practice of the invading forest farmers. Thus far 3.7 million hectares have been consumed by fire on the island, with the extermination, beyond doubt, of unknown species in local biomes that will never recover their pristine variety.

This volume is based on a symposium organized by the New York City Advisory Committee for the 150th meeting of the American Association for the Advancement of Science. A significant role of the AAAS is to bring together work from different scientific disciplines in a public meeting. Here the work is developed in its relevance to human welfare. The expectation is that larger numbers of reasonable people placed in possession of the same body of rational understanding will concert their opinion in the formation of wiser public policy.

Gerard Piel
President (1985-1986)
American Association for the
Advancement of Science

(1)   Woodwell, George M. The Carbon Dioxide Question. Scientific American Vol. 238, No. 1, p. 34 (January 1978)

Acknowledgments

The editor especially thanks Rosemary Lawlor for typing many drafts of this volume including the final camera-ready copy, the various people who reviewed the manuscripts included in this volume, David Savold of AAAS for many helpful comments and editorial suggestions, and Gerard Piel and Frances Maroncelli for help with the organization of the symposium upon which this book was based.

Ghillean T. Prance

1. Introduction to Tropical Rain Forests

ABSTRACT

There are many reasons for concern about the disappearance of the rain forests of the world. Before considering some of the implications of deforestation and its effects on the atmosphere described in the following chapters, this introduction reminds us of some of the important features of rain forests. The species diversity of rain forests and the interactions between different organisms are emphasized. Present knowledge of rain forest ecosystems does not even include a complete inventory and in many places species are being lost before they have ever been described. Rain forests are a source of many plants of economic potential, much of which has not yet been realized. There is a great need for further work on the economic botany of rain forests and on the study of the fast-disappearing indigenous people's knowledge of rain forest ecology and management.

INTRODUCTION

The purpose of this volume is to draw attention to the need for concern about tropical deforestation in terms of its effect on global atmosphere. However, before we turn to some of the factors which cause the rain forests to influence the atmosphere, climate, and rainfall, we should briefly review some of the basic biological facts about rain forests and also the causes of deforestation. I will discuss here the biology of the rain forests and the next chapter by Norman Myers will take up the issue of deforestation. With an understanding of these facts we can then make a more detailed study of some of the atmospheric problems addressed in the other chapters. I want to present some facts about rain forests to remind us of the diversity, the complexity of interaction, and the vast unrealized potential of economic products from this biome, taking examples mainly from the Amazon rain forest with which I am most familiar.

This volume is largely concerned with the tropical rain forests and tropical moist forests of the world which occur in the intertropical region in all places where the climate, topography, and soil allow. Hence the largest are in Amazonia, west central Africa in the Zaire Basin, and in Indonesia. Generally tropical moist forests occur in lowland regions with over 2000 mm of annual rainfall. However, it is not just the quantity of rain but also the seasonality that is a very important factor in determining the actual vegetation cover of any part of the tropics. When the monthly rainfall drops below 120 mm for longer than a month, tropical rain forest tends to be replaced by tropical moist forest. There is great variation in annual rainfall with the tropical rain forest areas varying from the extremely wet forests of the Chocó with over 9000 mm that are probably the wettest rain forests of the world (Gentry, 1982b) to the rather seasonal rain forests of Central Amazonian Brazil with as little as 1800 mm. When there is a strong seasonality, the trees become deciduous and so this is tropical deciduous forest rather than tropical rain forest. Because of its seasonal growth pattern, the deciduous forest has both less biomass and fewer species than tropical moist forest. Tropical moist forest lies between these two extremes of rain forest and deciduous forest in areas that are moderately seasonal and have less than 2000 mm of rain rather evenly distributed through the year. It is these species diverse rain forests (Fig. 1.1) and moist forests that are the focus of attention here, because of both their interest as species rich habitats and their rapidly accelerating destruction.

DIVERSITY

Diversity is a much used term in most descriptions of rain forests. This is because many (but not all) tropical rain forests are extremely diverse in terms of species number. They are the most species rich biomes in the world and are therefore of great value purely because of the genetic diversity they contain. However, the diversity extends to many other areas especially to the diversity of habitats that are available within the rain forest biome and the different vegetation types that come within the term tropical rain forest.

In recent years there have been many more detailed inventories of rain forest areas carried out by botanists who have taken the trouble to make accurate identification of all individual trees growing in their study areas. Consequently we now have a much better idea about species diversity than those based on forest inventories which relied on local names. To give a few examples: A hectare of forest which we inventoried near Manaus, Brazil (Prance et al., 1976) contained 179 species of trees over 15 cm in diameter and 236 species when diameter class was reduced to 5 cm. Gentry (1982a) found 258 species of trees (including all small ones) and lianas in only 1000 square meters, near Tutunendo in the Chocó region of Colombia. The Río Palenque Biological Station in the Pacific coastal forest of Ecuador contains 1033 species in an area of 1.7 square kilometers (Dodson and Gentry, 1978). The flora of Barro Colorado Island in the Panama Canal contains 1318 species in an area of 15 square kilometers (Croat, 1979). On that small island a plot of 50 hectares studied in detail by Hubbell and Foster (1983) contained 186 species of trees of 10 cm diameter or more. Gentry (1982a) gave an interesting comparison of forest inventory data from many sites and showed that the species diversity is quite varied from one rain forest area to another. He found that there is a strong correlation between species diversity and rainfall. The greater the rainfall (up to 4000 mm) the greater the species diversity. These few figures serve to demonstrate the diversity of tropical forests.

A common belief is that the Amazon forest is one uniform mass of forest, but in actuality it is a complex mosaic of different vegetation formations. We now have a better idea of this through the publications of the RADAMBRASIL (1972-82) survey of the region. Only slightly more than 50% of the region is covered by the typical rain forests on terra firme. The rest is a vast number of other vegetation types such as Amazon caatinga on sandy soils, lower montane forest, floodplain várzea and igapó inundated forest formations, flooded savannas, cerrado, and other savanna types, and various aquatic habitats. For a review of Amazonian vegetation types see Pires and Prance (1985). Unfortunately another increasingly common formation is the vast amount of secondary forest (called capoeira in Brazil) that occurs in areas that have been cut and used for a short time and then abandoned.

The terra firme rain forests are extremely diverse and more and more detailed quantitative inventories, such as those cited above, are showing the local variation dependent on local rainfall patterns, soil, topography, and altitude. There is much clustering of species in some areas, while in others species are much more randomly distributed (see Hubbell and Foster, 1983).

Even the seasonally flooded forest is quite variable depending on water type (black, white, or clear water) and length of the annual inundation.

These generalizations for Amazonia are true for other rain forest areas of the world which are also a mosaic of vegetation types. For example, within the rain forests of Sarawak on the island of Borneo occur patches of heath forest on white sand that closely resemble the white sand campina forests of central Amazonia. In Africa similar transition zones of semideciduous and deciduous forest border many of the rain forest areas. Smaller patches of rain forests occur around the tropics in places where the climate is suitable such as in humid parts of Central America and southern Mexico, some of the larger islands of the Caribbean, Hawaii, Madagascar, southwest Asia, etc.

INTERACTIONS

One of the most fascinating aspects of my studies of the Amazon forest has been that of various biological interactions such as pollination and seed dispersal. The study of these and other plant animal interactions furnishes us with the data we need to demonstrate that a forest is one huge ecosystem with the different organisms all linked together in a delicate network. I will give one example of a plant of considerable economic importance to the Amazon region, the Brazil nut (Bertholletia excelsa Humb. and Bonpl., Fig 1.2).

My colleague, Dr. Scott Mori, and I have now carried out extensive studies on the pollination of the Lecythidaceae, the Brazil nut, and its relatives. The Brazil nut flowers need to be cross-pollinated with pollen from another tree in order to set fruit. The flowers are of an extremely complex structure with the androecium extended laterally into a hood that covers the ring of fertile stamens around its base. Nectar is secreted in the hood, and the pollinator must life the hood to obtain access to the nectar. The flowers are visited by large bees, mainly of the subfamily Euglossineae, but also carpenter bees (Xylocopa). These bees land on the hood and enter the crack between the hood and its base. While they forage for nectar in the hood, their backs rub against the fertile stamens of the ring and they become heavily dusted with pollen which they carry from flower to flower. The Brazil nut flowers for about a one month period in November and provides good sustenance for the bees during that time. Our phenological studies, and those carried out by scientists of INPA at the Reserva Ducke near Manaus, have shown that the pollinators of the Brazil nut also visit a whole series of other species. Many of these species reach flowering peaks at slightly different times from the Brazil nut and thus help to feed the pollinators over a longer period. The Brazil nut will therefore not produce its fruit without the bees nor without its related species to provide bee food at other times.

The male Euglossine bees are also dependent on some of the epiphytic orchids in the forest because they visit these and gather their fragrances. They pack the scents in their hind legs and fly off to form a lek, a group that attracts the females for mating to take place. Therefore, if the orchids are not present in the forest the Brazil nut will also not produce. We have a large network of organisms that are interrelated through their interactions.

In addition, the Brazil nut falls to the forest floor where the outer shells are eaten by agoutis. The agoutis bury the seeds (scatter hoarding) and forget some of their caches thereby causing dispersal of the seeds. The survival of the Brazil nuts is therefore dependent on this large rodent on the forest floor as well as the bees in the canopy.

Because of the laws which prohibit the felling of Brazil nut trees, lone trees left standing in felled areas are often seen. This will not help because the pollinators are not present to enable production of the fruit. If the Brazil nut is to be grown in plantations they must be adjacent to areas where the Euglossine pollinators can survive.

One could list hundreds of known examples of intricately coevolved interactions between the different organisms of the forest. For example: insects that visit extrafloral nectaries and consequently defend the nectar-producing plant from other predators; trees such as Tachigali and Triplaris that are inhabited by protective fire ants; mycorrhizal fungi which help the tree to recycle nutrients efficiently (see Janos, 1983; St. John, 1985); cacique birds whose chicks are protected from botflies by hornets who build their nests in the same tree (see Smith, 1968); and the numerous pollination and dispersal examples that have been studied (see Prance, 1985). The destruction of the forest is breaking these interdependent relationships and, consequently, the whole structure of the ecosystem.

USEFUL PLANTS

Many well-known plants of the tropical rain forests have economic potential, such as rubber, cacao, peach palm, rosewood, and guaraná from the Amazon, and rattan, mangoes, and dipterocarp timbers from Asia. There are many more plants used by just a few Indians and other indigenous peoples that could be applied much more widely, and there are many others whose economic potential has not been realized at all. Many potential new foods, fibers, drugs, and sources of fuel will certainly become extinct before long if the present trend of destruction continues. In addition, some of these plants can probably be grown on a sustained yield basis without the total destruction of the forest presently taking place to create cattle pastures and other non-forest ecosystems.

One of the best sources of new information is from the fast-disappearing Amazonian Indians who have lived with the forest for several thousand years. The study of the ethnobotany of the plants used by indigenous peoples should be given the highest priority. It is a rewarding and useful experience to sit under the tutelage of an Indian teacher as he or she explains the use for each of the plants around and demonstrates knowledge of their ecology and management.

A recent study of the Chacobo Indians in the Bolivian Amazon by Brian Boom showed how truly dependent they are upon the forest plants, and what an extensive knowledge they have about uses for the diversity of species that surrounds them (Boom, 1985 a, b). In a sample hectare of rain forest in Chacobo territory, it was found that the Indians use 82% of the species (75 of the 91 tree species) and an amazing 95% of the individual trees in the forest (619 of the 649). These uses include all categories such as medicines, building materials, fuel wood, ornaments, paints, fibers, etc. Other studies are now underway to compare these data with those for other tribes, and a similar diversity of uses is being found for the Kaapor Indians of Maranhão, Brazil (William Balée, 1985).

The Amazon plant species that are presently used by more than a few Indians from the forest are minimal in comparison to the possibilities. Already commercial interest in some useful plants such as the palms pupunha (Bactris gasipaes) and babassu (Orbignya sp.), the moraceous tree mapati (Pourouma cecropiifolia), and the edible species of Calathea (Marantaceae-batata ariá) has developed. There are many more which could eventually come into greater use such as cupa (Cissus gongyloides in the Vitaceae), which is the source of carbohydrate for a few groups of Indians in the southern part of the State of Pará; bekú (Curarea tecunarum in the Menispermaceae), which is the contraceptive of the Dení Indians; or some of the 26 species of edible fungi that are eaten by the Yanomamo. Walking through the forest with an Indian, one is amazed by the number of different species for which he or she has use. Is this wealth of information going to be recorded or are we going to allow it to be lost forever?

Many of the plants which we already use have their wild relatives in the forest. It is essential to preserve these species. With modern genetic engineering they will become increasingly important when it becomes easier to take desirable characteristics such as disease resistance or higher protein content from wild species and place them into the cultivated ones. Thus the more than 20 wild species of Theobroma (Fig. 1.3) are important for the future of cacao and the 10 wild species of Hevea are important for further development of rubber. These are the species that are threatened by destruction of the forest before they are even properly known. For example, within the last 2 years I collected a new species of Hevea, the rubber genus, in the northwestern Amazon.

We can see that there are many compelling reasons for interest in, and preservation of, the remaining tropical rain forests of the world. They represent an incredible amount of genetic diversity of species that have countless different potential uses. The forest itself is a rich heritage which this generation should be able to pass on to the next because of its own value, and not only for its importance to the atmosphere and the climate which we will learn about in the other chapters in this volume (Fig 1.4).

REFERENCES

Balée, W., 1985: Personal Communication.

Boom, B., 1985a: Amazon Indians and the forest environment. Nature 314, 324.

_____, 1985b: 'Advocacy botany' for the Neotropics. Garden, 9 (3), 24-32.

Croat, T., 1979: Flora of Barro Colorado Island. Stanford University Press.

Dodson, C. and A. Gentry, 1978: Flora of the Río Palenque Science Center, Selbyana 4, 1-628.

Gentry, A. H., 1982a: Patterns of Neotropical plant species diversity. Evol. Biol. 15, 1-84.

Gentry, A. H., 1982b: pages 112-136 In: G. T. Prance (ed.). Biological diversification in the tropics. Columbia University Press, New York.

Hubbell, S. P. and R. B. Foster, 1983: Diversity of canopy trees in a neotropical forest and implications for conservation. Pages 25-41. In: S. L. Sutton, T. C. Whitmore and A. C. Chadwick (eds.). Tropical rain forest: ecology and management. Spec. Publ. Brit. Ecol., Soc. 2, Blackwell, Oxford.

Janos, D., 1983: Tropical mycorrhizas, nutrient cycle and plant growth. Pages 327-345. In: S. L. Sutton, T. C. Whitmore and A. C. Chadwick (eds.). Tropical rain forest: ecology and management. Spec. Publ. Brit. Ecol. Soc. 2. Blackwell, Oxford.

Pires, J. M. and G. T. Prance, 1985: The vegetation types of the Brazilian Amazon. Pages 109-145. In: G. T. Prance and T. E. Lovejoy. Key Environments: Amazonia, Pergamon Press, Oxford.

Prance, G. T., 1985: The Pollination of Amazonian plants. Pages 166-191. In: G. T. Prance and T. E. Lovejoy (eds.). Key Environments: Amazonia. Pergamon Press, Oxford.

Prance, G. T., W. A. Rodrigues and M. F. da Silva, 1976: Inventario florestal de um hectare de mata de terra firme km 30 da Estrada Manaus-Itacoatiara, Acta Amazonica 6, 9-35.

Projeto RADAMBRASIL, 1971-1981: Levantamento de Recursos Naturais Vols. 1-22. Ministerio das Minas e Energia.

St. John, T., 1985: Mycorrhizae. Pages 277-283. In: G. T. Prance and T. E. Lovejoy. Key Environments: Amazonia. Pergamon, Oxford.

Smith, N. G., 1968: On the advantages of being parasitized. Nature 219, 690-694.

2. Tropical Forests: Patterns of Depletion

INTRODUCTION

Tropical forests are being depleted, both quantitatively and qualitatively, at accelerating rates. If present land-use trends and exploitation patterns persist (and they are likely to accelerate), large sectors of tropical forests will become markedly modified, if not fundamentally transformed or eliminated outright, within the next three to five decades. In some parts of the biome, many forests will have been reduced to degraded remnants, if not destroyed, by the end of this century.

This much is not in doubt: tropical forests are declining. All authoritative observers agree that a pattern of disruption and destruction is overtaking the biome and resulting in complete and permanent elimination of extensive tracts of forest, and in significant disturbance through degradation and impoverishment of forest ecosystems to a degree that will not permit recovery to complete primary status within less than several decades, in some cases even centuries.

The key question is how fast are these depletive processes affecting remaining forests? Is the principal injury to tropical forest ecosystems of a sort that stops short of outright deforestation, i.e. imposing considerable injury to forest ecosystems while leaving some trees standing? Or is it more a case of total elimination of forest cover, with the land being given over to an entirely different purpose?

It is centrally important to draw a clear distinction between these two forms of depletion. Much of the confusion in recent years about supposed discrepancies between different estimates of depletion rates has arisen through misunderstandings concerning the nature of depletion or of "conversion," to use a catch-all term that stands for all types of forest depletion, ranging from marked modification to fundamental transformation to outright destruction. For the purposes of this chapter, the first set of processes, viz. major disturbance, is considered to result in qualitative depletion, leading to pronounced (as opposed to marginal) impoverishment of forest ecosystems. The second set of processes is viewed as deforestation, i.e. complete and permanent removal of all forest cover.

RATES OF DEPLETION

Two surveys have been conducted in recent years, these being the first and only surveys to draw in major measure on the findings of remote-sensing technology, viz. a technology that supplies objective and systematized data. The first of these surveys was commissioned by the National Academy of Sciences, and its results were published by the National Research Council (Myers, 1980; for an up-date, see Myers, 1984). The second survey was conducted by the Food and Agriculture Organization (FAO), in conjunction with its sister agency, the United Nations Environment Program (UNEP), and the results published in 1982 (FAO and UNEP, 1982). The first of these surveys looked both at deforestation and disruption, the second confined itself to deforestation. The first survey produced a depletion estimate of 200,000 square km a year, out of a biome comprising rather less than 10 million square km. A rough breakdown, presented in preliminary and approximate terms, suggested that about half of the 200,000 square km could be categorized as deforestation, and the other half as disruption. The second of the surveys has asserted a depletion rate of 76,000 square km a year, this being a figure confined to deforestation.

Following publication of these two reports, both sets of authors have conducted detailed analyses of their findings to see whether reconciliation, however partial, can be made of the two sets of figures for deforestation, viz. around 100,000 square km a year and 76,000 square km. We find that when we "compare apples with apples," i.e. when we confine out attentions to tropical moist forests below 1500 m (and discount montane forests, bamboo forests, drier forests that border on woodlands, and the like), the deforestation rates can be accurately set at 91,000 square km and 73,000 square km respectively—no great discrepancy (Hadley and Lanly, 1983; Houghton et al 1985; Melillo et al., 1984; Myers, 1985). The analysis further indicates that disruption can be reliably estimated to be accounting for some 100,000 square km a year, for a depletion total of slightly less than the 200,000 square km originally proposed.

MAIN AGENTS OF DEPLETION

There are four main agents involved. The first is the commercial logger, who now accounts for some 45,000 square km of primary forest each year. His impact is not significant in the broad picture, however, because most logged forests are soon occupied by smallscale cultivators and other farmers. The fuelwood gatherer, who is much more active in open woodlands and thorn scrub areas, accounts for about 25,000 square km of forest each year. The cattle raiser, confined to Latin America, is responsible for at least 20,000 square km. By far the most important agent of all, the smallscale forest farmer--operating variously as shifting cultivator, migrant squatter, or landless "colonist"--is depleting as much as 150,000 square km a year, possibly rather more. These figures round out to a total of 200,000 square km a year, or about 2% of the remaining biome.

Of course--and this emphasizes a central factor--these figures represent no more than informed estimates. There are many problems with surveys of tropical forests, notably the diverse definitions and classifications used by the several dozen governments of the biome, the inaccuracy of many data presented, and an occasional political compulsion to "fudge" the figures. Thus the estimates set out above are certainly not presented as a concise account of what is definitely occurring in tropical forests. The statistical details need to be treated with a range of reliability. Indeed many quantified statements should be qualified with phrases such as "so far as can be ascertained..." or "best-judgment assessments indicate that...." In other words, one needs to aim, on the one hand, to avoid bogus accuracy, and on the other hand, to describe semi-documented situations with an appropriate degree of "precise imprecision."

DIFFERENTIATED RATES OF DEPLETION

Plainly some sectors of the biome are being hit harder

than others.   In fact, depletion patterns turn out to be highly differentiated.   Certain sectors of the biome are undergoing   widespread depletion at rapid rates, other sectors are experiencing moderate depletion at intermediate rates, while still other areas are encountering little change.   In brief, we can say that virtually all lowland forests of the Philippines and Peninsular Malaysia seem likely to become heavily logged by 1990, or very shortly thereafter.   Much the same applies to most parts of West Africa.   Little   could remain of Central America's forests by 1990.   Almost all of Indonesia's lowland forests have been scheduled for timber exploitation by the year 2000, and at least half by 1990. Extensive portions of Amazonia in Colombia and Peru could be claimed for cattle ranching and various forms of cultivator settlement by the end of the century; and something similar is true for southern and eastern sectors of Brazilian Amazonia.

By contrast, Central Africa is sparsely inhabited and possesses abundant minerals.   This reduces the incentive for governments to liquidate their forest capital in order to supply funding for various forms of economic development.   Hence there could well remain large expanses of little-disturbed forest in Central Africa at the turn of the century.   Similarly, the western portion of Brazil's Amazonia, because of its remoteness and perhumid climate, could undergo only moderate change.

In sum, the overall outcome is likely to be extremely patchy, both in terms of geographic areas and degrees of depletion.   For a more detailed account of differentiated depletion patterns, see Appendix I.

FUTURE PROSPECTS

Since the forest farmer is by far the most important agent of depletion, let us briefly consider his prospect for the future.   Already around 200 million people gain their subsistence through this type of agriculture (according to some experts, e.g. Denevan, 1982; UNESCO, 1978; and Watters, 1971, the total number of forest farmers could have reached 200 million as early as 1970, and could now be approaching 300 million--although the FAO/UNEP Report, 1982, appears to propose a total of well under 100 million, albeit a total documented in less-than-consistent detail).   The greatest loss to forest farmers is currently thought to be occurring in Southeast Asia, where cultivators clear a minimum of 85,000 square km a year, adding to the 1.2 million square km of formerly forested croplands in the regions (Chandrasekharan, 1978; Kartawinata et al., 1981; Myers, 1980 and 1984). Africa south of the Sahara is believed to have lost one million square km of forest to these cultivators even before modern development started to gain momentum after World War II; by the mid-1970s the region's loss has been estimated to have reached 40,000 square km a year, in a zone of 400,000 square km of forestlands under that style of agriculture (Braun, 1974; Hauck, 1974; Persson, 1977). Latin America is likewise believed to have been losing forest to smallscale cultivators at a rate of 40,000 square km a year by the mid-1970s, though fewer details are available (Denevan, 1982; Watters, 1971).

But on-the-ground investigations reveal that census figures for forest farmers in several countries are incomplete. For various political reasons, governments prefer to ignore refugees, guerillas and other insurgents, and general "trouble makers." Thus official estimates for the Philippines, Thailand, and half a dozen other countries could well be doubled or even tripled to make a realistic figure; while in Ivory Coast and several other countries of West Africa, official estimates deserve to be expanded by at least one half to take account of illegal immigrants fleeing from the Sahel disaster (Myers, 1984).

Precisely because of migration patterns on the part of numbers of landless peasants who believe they can find salvation for themselves in the only other "free," i.e. public, lands left available in their countries, the numbers of forest farmers can be expected to continue to grow at a rate far faster than that of national populations. Whereas population growth in the countries concerned is projected to produce an increase of almost 50% during the last two decades of this century, the number of forest farmers could well double. As for the longer-term future, none of the major tropical forest countries is projected to reach zero population growth, even with stepped-up family planning programs, until late next century, if not until early in the 22nd century, due to the "youthfulness" of their present populations. So we must anticipate a future where the population of the Philippines grows from its 1980 total of 49.1 million to 125 million; for Brazil from 118.7 million to 281 million; and of coastal West Africa (not counting Nigeria, with its 85 million projected to reach 459 million) from 46 million to 263 million. Of the incremental numbers, a solid proportion will, if current land-use trends persist, become forest farmers, to add to the existing multitudes. At the same time, however, let us not forget a hopeful aspect: Zaire, Gabon, and Congo, with a total forest area of 1.2 million square km are projected to grow from a 1980 total of around 35 million to an eventual total of no more than (sic) 168 million. For further details, see table 2.1.

REMOTE SENSING

In a situation where information is of varying quality, we are fortunate to call upon a new form of technology that helps us in assessing tropical forest resources. It is remote sensing. The merits of this technology are that it is systematic and objective. Through satellite imagery, we can make inventories of large portions of the earth's surface in a short space of time, often at remarkably low cost. At the time of the National Academy study in 1979, remote-sensing findings were available for Thailand, the Philippines, Brazil, Nigeria, and several other countries, totaling almost half the biome. In more cases than not, satellite imagery revealed that countries actually possessed less forest cover than they believed they did.

As author of the National Academy study, I had to depend on the limited capacities of LANDSAT 3, which produced "images" covering blocks of land 185 km on each side, with each of four bands producing around seven million picture elements ("pixels") or minimal-area data points in each image. Numerous as these data points sound, they do not offer nearly enough resolution to indicate just what is going on at ground level. Worse still, they cannot differentiate between the vegetation of a forest and the vegetation of a cropland patch, hence they cannot discern the impact of forest farmers.

Fortunately the United States has recently lofted a more advanced satellite, LANDSAT 4. One of the sensors of this satellite, the thematic mapper, offers a resolution two and a half times the "detail documentation" power of its predecessor (30m x 30m), so it can engage in fine-grain mapping of forest features. In fact, LANDSAT 4 is sensitive enough to take pictures the size of a tennis court 650 km below, and often to do so through partial cloud and darkness. Instruments on board pick up these varied signals from earth, and interpret them to indicate whether they derive from rock, soil, water, or vegetation--and, most importantly, vegetation of different

Table 2.1
Population Projections for Some Tropical Forest Countries

Country	Population in 1980 (In Millions)	Rural Population (%)	Population in 2000 (In Millions)	Year of Net Reproduction Rate	Year of Stationary Population	Total of Stationary Population (In Millions)
Brazil	118.7	39	176.5	2010-15	2075	281
Colombia	26.7	40	39.4	2005-10	2065	60.2
Costa Rica	2.2	59	3.3	2000-05	2065	4.8
Ecuador	8	58	13.6	2020-25	2085	28.4
Guatemala	7.2	64	12	2020-25	2085	24.2
Guyana	0.8	60	1.2	2000-05	2065	1.8
Honduras	3.7	69	6.7	2025-30	2090	16
Mexico	69.8	36	115	2010-15	2075	202.6
Nicaragua	2.6	51	4.7	2025-30	2090	10.8
Panama	1.8	50	2.8	2005-10	2070	4.3
Peru	17.4	38	27.3	2015-20	2080	48.8
Surinam	0.4	34	0.6	2005-10	2070	0.9
Venezuela	14.9	25	23.8	2005-10	2075	38.6

TABLE 2.1 (continued)
Population Projections for Some Tropical Forest Countries

Country	Population in 1980 (In Millions)	Rural Population (%)	Population in 2000 (In Millions)	Year of Net Reproduction Rate	Year of Stationary Population	Total Stationary Population (In Millions)
Bangladesh	88.5	91	141	2030-35	2125	388.2
Burma	34.8	78	54	2025-30	2095	89.7
India	673.2	78	994.1	2015-20	2115	1,621.5
Indonesia	146.6	82	216	2015-20	2110	388.4
Malaysia	13.9	73	20.8	2000-10	2070	30
Papua New Guinea	3.1	87	4.4	2030-35	2125	9.4
Philippines	49.1	68	76.9	2010-15	2075	125
Thailand	46.9	87	68	2000-05	2070	102.5
Vietnam	54.2	78	87.9	2010-15	2075	153.4
Cameroon	8.5	71	14.2	2035-40	2110	40.8
Congo	1.5	60	2.7	2035-40	2100	10.5
Gabon	0.7	68	0.9	2035-40	2130	2
Ivory Coast	8.2	68	14.8	2035-40	2110	47.4
Madagascar	8.7	84	16.1	2035-40	2110	51.1
Zaire	28.3	70	51	2035-40	2110	156.1

Source: M. T. Vu and A. Elwan, 1982, "Short-Term Population Projection 1980-2000, and Long-Term Projection 2000 to Stationary Stage, by Age and Sex for All Countries of the World," The World Bank, Washington, D.C.

sorts. The images are then digitized and transmitted to earth-based receiving stations where they are converted to color or black/white pictures that are remarkably revealing. During the course of a photographic cycle of 18 days, within 14 circuits of the globe a day, the satellite can pass over virtually every part of the earth.

LANDSAT 4 will mark a quantum advance in our ability to assess the tropical forest situation. LANDSAT 3 was not able to pick out small plots of forest clearings, which helps explain why FAO and UNEP, with their emphasis on remote-sensing information, gives little emphasis in their reports to forest farmers--centrally significant as these farmers are to our entire understanding of forest depletion patterns and trends. We can look forward to these problems being resolved now that we are accumulating information from LANDSAT 4 (as long as the satellite remains in good working order). Within just a few years we should be able to come up with solid evidence of what is truly happening to tropical forests.

CONCLUSION

Thus, the decline of tropical forests. The situation represents a travesty, if not a tragedy, of development and conservation. It is all the more regrettable in that there are plenty of sound reasons why we should view tropical forests as a source of many goods and services to support our welfare into the indefinite future. Were we to undertake a comprehensive evaluation of all that tropical forests have to offer us, we would surely find it in our interests to reverse the present tendency towards elimination of these valuable and unique stocks of natural resources.

REFERENCES

Braun, H., 1974: Shifting Cultivation in Africa. Report of the FAO/SIDA/ARCN Regional Seminar of Shifting Cultivation and Soil Conservation in Africa. Food and Agriculture Organization, Rome, Italy.

Chandrasekharan, C. (editor), 1978: Shifting cultivation. Forest News 2 (2), 1-25.

Denevan, W. M., 1982: Deforestation in Latin America. Paper prepared for Office of Technology Assessment, Washington, D.C. Department of

Geography, University of Wisconsin, Madison, Wisconsin.

Food and Agriculture Organization and United Nations Environment Programme, 1982: Tropical Forest Resources, three volumes. Food and Agriculture Organization, Rome, Italy, and United Nations Environment Programme, Nairobi, Kenya.

Hadley, M. and J. P. Lanly, 1983: Tropical Forest Ecosystems: Identifying Differences, Seeing Similarities. Nature and Resources 19 (1), 2-19.

Hauck, F. W., 1974: Shifting cultivation and soil conservation in Africa. Soils Bulletin No. 24. Food and Agriculture Organization, Rome, Italy.

Houghton, R. A. et al., 1985: Net Flux of Carbon Dioxide from Tropical Forests in 1980. Nature 316: 617-620.

Kartawinata, K. et al., 1981: The impact of man on a tropical forest in Indonesia. Ambio 10 (2-3), 115-119.

Melillo, J. M. et al., 1985: A Comparison of Two Recent Estimates of Disturbance in Tropical Forests. Environmental Conservation 12: 37-40.

Myers, N., 1980: Conversion of Tropical Moist Forests. National Research Council, Washington, DC.

Myers, N., 1984: The Primary Source W. W. Norton, New York.

Persson, R., 1977: Forest Resources of Africa, Part II: Regional Analysis. Department of Forest survey, research notes no. 22. Royal College of Forestry, Stockholm, Sweden.

UNESCO, 1978: Tropical Forest Ecosystems. UNESCO, Paris, France.

Watters, R. F., 1971: Shifting Cultivation in Latin America Food and Agriculture Organization, Rome, Italy.

APPENDIX I: Country-by-Country Review of Depletion in Tropical Forests

This summary review is taken from the 1980 publication, Conversion of Tropical Moist Forests (Myers, 1980). The author sees no reason to change any of the details now that we have reached 1985.

A. Areas Undergoing Broad-Scale Depletion at Rapid Rates

1. Most of Australia's lowland tropical forests, both

rain forests and seasonal forests, due to timber exploitation and planned agriculture; could be little left by 1990 if not earlier.

2. Most of Bangladesh's forests, both lowland and upland, predominantly rain forests, due to timber exploitation, forest farming, and population pressure; could be little left by 1990 if not earlier.

3. Much of India's forests, predominantly seasonal forests, mainly upland, due to forest farming and population pressure; could be little left by 1990.

4. Much if not most of Indonesia's lowland forest, predominantly rain forests, due to timber exploitation, forest farming, and transmigration programs; could be little left in Sumatra and Sulawesi by 1990, in Kalimantan and most of the smaller islands by 1995, and in Irian Jaya by the year 2000.

5. Much of Sumatra's and Sabah's lowland forests, almost all rain forests, due to timber exploitation; could be little left by the year 2000 if not earlier.

6. Most of Peninsular Malaysia's lowland forests, almost all rain forests, due to timber exploitation and planned agriculture; could be little left by 1990 if not earlier.

7. Much if not most of Melanesia's lowland forests, due to timber exploitation and planned agriculture; could be little left by 1990.

8. Most of the Philippines' lowland forests, predominantly rain forests, because of timber exploitation and forest farming; could be little left by 1990 if not earlier.

9. Much if not most of Sri Lanka's forests, predominantly rain forests, mostly upland, due to timber exploitation and forest farming; could be little left by 1990.

10. Much if not most of Thailand's forests, almost all seasonal forests, both lowland and upland, due to timber exploitation (especially illegal felling) and forest farming; could be little left by 1990 if not earlier.

11. Much of Vietnam's forests, almost all seasonal forests, both lowland and upland, especially in the south, due to forest farming, timber exploitation and immigration from the north; could be little left by 1990.

12. Parts of Brazil's southern and eastern sectors of Amazonia, lowland rain forests, notably in Rondônia, Pará and Mato Grosso; due to cattle raising and colonist settlement; appreciable tracts could be depleted by 1990 and much more by the year 2000.

13. Most if not virtually all of Brazil's Atlantic coast strip of moist forest, due to timber exploitation and cash-crop agriculture, notably sugarcane plantations; could be little left by 1990 if not a good deal earlier.

14. Much if not most of Central America's forests, notable rain forests, both lowland and upland, due to forest farming, cattle raising, and timber exploitation; could be little left by 1990 if not earlier.

15. Parts of Colombia's lowland rain forests on the borders of Amazonia, especially in Caqueta and Putumayo, due to colonist settlement and cattle raising; extensive tracts could depleted by 1990, and much more by the year 2000.

16. Much of Ecuador's Pacific coast forests, mostly very wet and very rich rain forests, both lowland and upland, due to plantation agriculture and some timber exploitation; could be almost entirely depleted by 1990.

17. Much if not most of Madagascar's forests, especially rain forests, both lowland and upland, due to forest farming and timber exploitation; could be little left by 1990 if not earlier.

18. Much if not most of East Africa's relict montane forests, especially in northern Tanzania, mostly seasonal forests, due to timber exploitation, firewood cutting, and forest farming; could be little left by 1990.

19. Much if not most of West Africa's forests, mainly seasonal forests, due to timber exploitation and forest farming; could be little left by 1990 if not earlier.

B. Areas Undergoing Moderate Depletion at Intermediate Rates

These areas cannot be so readily listed as those under Areas Undergoing Broadscale Depletion at Rapid Rates, since less is known about their present status and future prospects. The listing is deliberately conservative.

1. Parts of Burma's lowland forests, almost all seasonal, due to forest farming and some timber exploitation; appreciable areas could be depleted by the year 2000 if not earlier.

2. Parts of Papua New Guinea's forests, mostly seasonal, both lowland and upland, due to timber exploitation and forest farming; extensive areas could be depleted by the year 2000 if not earlier.

3. Parts of Brazil's Amazonia forests, lowland rain forests, notable in Amapá, Acre, sections of Trans-Amazon

Highway system and of the várzea floodplains, and areas selected for timber exploitation, eg. Tapajós River area; due to colonist settlement, forest farming, cattle raising, and timber exploitation; appreciable tracts could be depleted by 1990.

4. Parts of Colombia's Pacific coast forests, very wet and very rich rain forests, both lowland and upland, due to timber exploitation; extensive sectors could become depleted by 1990.

5. Much of Ecuador's Amazonia forests, almost all rain forests, both lowland and upland, due to colonist settlement, forest farming, some planned agriculture, and also oil exploitation; appreciable areas could be depleted by 1990, and much more by the year 2000.

6. Much of Peru's Amazonia forest, almost all rain forests, both lowland and upland, due to colonist settlement, forest farming, and some planned agriculture; appreciable areas could be depleted by 1990, and much more by the year 2000.

7. Parts of Cameroon's forests, both seasonal and rain forests, both lowland and upland, due to timber exploitation and forest farming; extensive areas could be depleted by the year 2000.

C. Areas Apparently Undergoing Little Change

Like Areas Undergoing Moderate Depletion, these areas cannot be so readily listed as those under Areas Rapidly Undergoing Broadscale Depletion, since less is known about their present status and future prospects. The listing is deliberately conservative, especially as concerns the long term.

1. Much of Brazil's western Amazonia, lowland rain forests, generally wetter and richer than eastern Amazonia; the except for some timber extraction in limited areas, and some cultivation of várzea floodplains, exploitation of this huge zone could prove difficult in view of its unusually wet climate and distance from markets; it is reasonable to anticipate—so far as can be ascertained, and the point is stressed—that much of this vast tract of lowland rain forest could remain little changed for some time, possibly until the year 2000, even for a time thereafter.

2. Much of the forests of French Guiana, Guyana, and Surinam, almost all rain forests, both lowland and upland;

timber exploitation, at present very limited, may expand, but, because population pressures are low, there is little likelihood of widespread colonist settlement and forest farming. So it is reasonable to anticipate--with caveat as under Brazil above--that large areas may remain little changed for a good while to come, possibly until the year 2000, conceivably for a time thereafter.

3. Much of the Zaire basin, comprising Congo, Gabon, and Zaire; some rain forest in Gabon, remainder mainly seasonal, almost entirely lowland; population pressures are low, and there are abundant mineral resources on which to base national economic development; timber exploitation, primarily limited to northern Congo and to Gabon, could expand; but in the main, it is reasonable to anticipate--with caveat as under Brazil above--that large areas may remain little changed for a good while to come, possibly until the year 2000, even for a time thereafter.

George M. Woodwell, Richard A. Houghton,
Thomas A. Stone

3. Deforestation in the Brazilian Amazon Basin Measured by Satellite Imagery

INTRODUCTION

The earth's remaining tropical forests are under extraordinary pressures for economic development (see Myers, Chapter 2 this volume) primarily in support of the expansion of agriculture. The forests are extensive in area, and many are magnificent in stature and in biotic diversity. Some are very old and most remain the source of support for indigenous people, as well as the source of diverse products, such as rubber, for the technologically developed world. The forests with their rivers and soils are the major reservoir of biotic diversity on earth. They are also large enough in area and in stature to contain sufficient carbon in plants and soils that, if released into the atmosphere as carbon dioxide, atmospheric concentration would increase appreciably. Information on changes in the area of tropical forests regionally and globally is now fundamental to management of the earth's resources. It may be fundamental to exerting human control over the temperature of the earth. How can information be developed on changes in the area of forests globally?

The greatest promise in both retrospective and current measurements of changes in area of forests lies in the use of satellite imagery (NASA, 1983; Woodwell et al., 1983; Woodwell (ed.), 1984; Klemas and Hardisky, 1983). LANDSAT 1,2,3,4, and 5 imagery is available from 1972 through the ensuing decade and will continue through 1985 and probably longer. The question is the extent to which LANDSAT, or other satellite imagery that has low resolution but wide coverage by comparison with aerial photography, can be used to measure changes in the area of forests. The problem has been addressed previously in forests of the temperate zone (Woodwell (ed.), 1984) with initial success. Can a

23

technique be developed for use in the tropics where experience on the ground may be limited or lacking?

One approach to the use of satellite imagery involves attempts at classification of the vegetation throughout the entire scene. The number of classes that can be recognized and identified with fidelity in LANDSAT imagery is normally limited to ten or less. Any numerical appraisal of change based on classification is subject to the errors inherent in any classification: there are questions about boundaries in addition to the fundamental decision about the class. A simpler technique that reduces or eliminates a reliance on classification would be desirable. One such technique involves the subtraction of one image from a second image to produce a new, third, residual image of the differences between the two originals. Subtraction is possible because the images exist as digital data in cellular units called picture elements or pixels. The advantage is that areas that have not changed are not considered and do not contribute errors due to misclassification. The images can be selected to span a period of time and to allow an estimate of rates of change. An emphasis on drastic change, easily identified in the images, such as that from forest to non-forest or from non-forest to forest simplifies the analysis further.

The technique is subject to errors. Great care must be taken in superimposing or registering images. An error of one or several pixels, about 0.45 ha, will produce spurious changes along margins of areas actually changed. There are simple ways of testing for and correcting misregistration. The second source of error arises from reliance on a large change in reflectivity as the basis of measurement. For example, a change in reflectivity on the ground that is less than a pixel in area may cause a sufficient brightening to flood the radiometer and produce an image in which that pixel is recorded as changed. Such an effect would be important along margins of areas that have changed where the areas changed are small but the ratio of margin to area is high. A correction may be appropriate in such circumstances to avoid exaggeration of the area changed. There is an advantage of course when an independent appraisal of such errors can be made for some small segment of the total area through the use of aerial photography or other higher resolution data.

Two satellite systems are useful in measurement changes in area of forests (Woodwell et al., in press; Tucker et al., 1983). LANDSAT provide 30 and 80m resolution from the Thematic Mapper (TM) and the

Multispectral Scanner (MSS); the NOAA AVHRR satellites provide 1.1 km resolution from portions of the spectrum that are effective in photosynthesis. LANDSAT data are taken in a swath 185 km wide every 14 to 18 days. The NOAA AVHRR satellites acquire data from the same area in a swath 2,700 km wide daily. Similar NOAA AVHRR satellites have been in operation since 1979. Global coverage is routinely obtained by LANDSAT. AVHRR coverage of specific regions must be requested in advance and those data are not commonly saved. We have used data from both these satellites in a preliminary analysis of deforestation in a portion of the Amazon Basin in Brazil.

Our first application of the technique in a tropical region was in the forests of Rondônia (Fig. 3.1). The total area of forest in Rondônia was 207,986 km^2 before colonization (IBGE, 1979). Rondônia is the focus of a large federal resettlement program that has attracted hundreds of thousands of the landless of southern Brazil, especially from the state of Paraná, where agriculture is becoming increasingly industrialized. Colonization of Rondônia began in 1970 with the establishment of the Brazilian Institute for Colonization and Reform of Agriculture (INCRA) and the designation of large portions of Rondônia for settlement. Prior to 1960 Rondônia was populated by scattered mineral prospectors, rubber gatherers, and perhaps 10,000 Amerindians. By 1980 the state had a population of 500,000 (World Bank, 1981). Between 1970 and 1980 the population of Rondônia increased at the rate of 16% per year. The area cleared by the settlers has increased at a much more rapid rate. Brazilian federal statistics show that the area cleared in 1975 was 1,200 km^2; in 1978 it was 4,200 km^2; and in 1980 the area was 7,600 km (Fearnside, 1984). These values have been criticized as an underestimate (Fearnside, 1982); they do not, of course, reflect the contemporary situation. We examined a portion (one LANDSAT scene or 32,000 km^2) of the area colonized in Rondônia to determine the area cleared and recent rates of clearing. We used the higher resolution data from LANDSAT to calibrate lower resolution data from NOAA AVHRR to estimate the area cleared for all Rondônia by 1982.

METHODS

LANDSAT data were purchased for an area in central Rondônia that an analysis of imagery from 1982 NOAA7 AVHRR

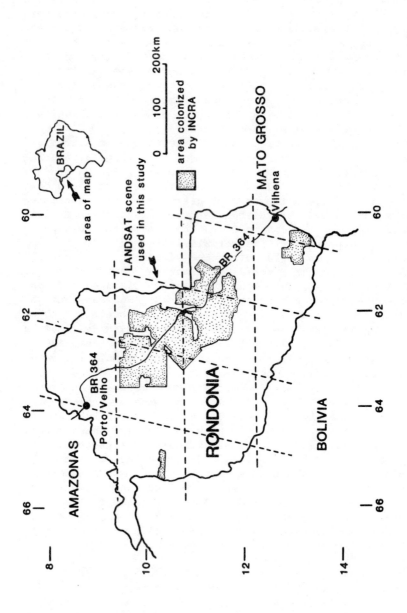

Fig. 3.1 The Brazilian state of Rondônia and the area covered by this study. Dashed lines indicate approximate boundaries of LANDSAT scenes (From Woodwell et al., in press).

showed to be in the process of being cleared. The AVHRR imagery had been obtained by Dr. C. J. Tucker of NASA's Goddard Space Flight Center (Tucker et al., 1983). A search of LANDSAT data from INPE, Brazil's space agency, and from the EROS Data Center in Sioux Falls, South Dakota provided several LANDSAT tapes between 1973 and 1981. Three tapes of high quality were used in the analyses: June 1976, August 1978, and May 1981 (Fig. 3.2a, b, c). Photographic images (1:500,000 scale) of the LANDSAT data were obtained with a copy of the July 1982 NOAA7 AVHRR data acquired by Dr. Tucker.

The three LANDSAT images were used in the analyses of changes summarized above (see also Woodwell et al. 1983). The major advantage of the method is that it emphasizes the changes that are of interest and ignores areas that have not changed. The errors of classification are almost eliminated. Computer time is saved and costs are lowered. The change detection technique is applicable to both LANDSAT and AVHRR data and, in fact, any digital image that examines the same portions of the earth's surface at more than one time.

The two LANDSAT multispectral scanner (MSS) bands used in this analysis were MSS5 (visible red) and MSS7 (near infra-red). The red part of the spectrum is absorbed by chlorophyll and the near infra-red portion is reflected strongly by healthy vegetation. Any significant change in land use that affects reflectivity, over an area larger than 0.45 ha (one pixel) will result in that pixel's being included in the difference image.

Pairs of LANDSAT images (185 x 185 km) were registered to one another in the computer to span periods of two years, three years, and five years (1976 to 1978, 1978 to 1981, and 1976 to 1981) for the analyses. Within each LANDSAT scene, five 30 x 30 km subscenes, each representing different rates of deforestation, were selected. The change detection algorithm was then applied to each of the subscenes and the best technique for the classification of change was chosen.

Preliminary determination of the area cleared in all of Rondônia using AVHRR data was by a simple comparison of the thermal band (Channel 3) data from the AVHRR with the LANDSAT data from the same area (Fig. 3.3). The grids of clearing were evident in the thermal band because roads and other cleared areas were warmer and therefore in contrast to the cooler surrounding forest. The cleared areas were warmer, presumably due to the loss of canopy, the reduced evapotranspiration and the increased soil temperature. A range of values of the thermal data that defined areas that had been cleared was selected. The area cleared determined by AVHRR within the LANDSAT scene was defined. This procedure yielded two estimates, LANDSAT- and AVHRR-based, of clearing within the LANDSAT scene.

Because the LANDSAT and AVHRR data were not from the same year, a direct comparison of the areas cleared could not be made. But estimates of the clearing in 1982 for the area of the LANDSAT scene were made by interpolating both linearly and exponentially from the LANDSAT data. Use of two estimates provided a range of estimates of area cleared by 1982 within the LANDSAT scene. A calibration factor between the two types of data was determined and the factor was used with the AVHRR data from the entire state of Rondônia to estimate the area cleared for all Rondônia.

LANDSAT ANALYSES

The change detection algorithm applied to all pixels within the three LANDSAT scenes of the same area (3,200,000 ha) yielded the following results:
The rate of deforestation for the two year period (t_1-t_2) 1976 to 1978 was 27,000 ha/yr (54,000 ha total).
The rate of deforestation for the three year period (t_2-t_3) 1978 to 1981 was 55,000 ha/yr (165,000 ha total).
The rate of deforestation for the five year period (t_1-t_3) 1976 to 1981 = 44,000 ha/yr (220,000 ha total).
Rates of clearing within the area of this one LANDSAT path-row doubled from the 1976-1978 period to the 1978-1981 period. Brazilian federal statistics for the entire state of Rondônia also reported a doubling in the rate of forest clearing. The rate reported for the entire state rose from 99,000 ha/yr (1975-1978) to 170,000 ha/yr (1978-1980) (Fearnside 1984). The method used here is internally consistent; it produced the same area of total deforestation (within 1%) by two different analyses (1976-1978) + (1978-1981)) = (1976-1981). The five year interval was sufficient to determine total clearing but not rates of clearing. Finally, a non-forest to forest transformation was not seen; presumably population pressure did not allow land to revert to forest.

We are not aware of any other direct estimates of forest clearing for 1982 for the entire state of

Rondônia, there are estimates of the total area cleared in Rondônia, however, for earlier years. To compare these estimates with the work reported here, we must (1) determine how much of the area within the LANDSAT scene was cleared prior to 1976 and (2) extend the analysis of the scene to the entire state. Both calculations involve classification of satellite imagery rather than change detection.

The area in the LANDSAT scene cleared prior to 1976 was estimated from more recent work with an adjacent scene with a similar area of cleared land in 1981. Classification of the scene suggested that 1,519 km^2 of forest had been cleared prior to 1978 in the adjacent scene (unpublished data). A reasonable minimum for the amount of clearing up to 1976 for the scene we are interested in would be 1,000 km^2. The total area cleared up to 1981 would then have been 2,200 km^2 plus 1,000 km^2 or 3,200 km^2, about 10% of the entire LANDSAT scene.

NOAA-AVHRR ANALYSES

The NOAA AVHRR thermal band (Channel 3) data for the same area provided an estimate that 3,677 km^2 had been cleared by July 1982. If we assume that LANDSAT data because of higher resolution are more accurate in determining the area cleared than the AVHRR data, a calibration factor can be developed between the AVHRR and LANDSAT data. Because of the difference in dates, we first projected the 1981 LANDSAT data to 1982. The area cleared in the 3,200,000 ha of the entire LANDSAT scene for 1982 was estimated by using both a linear rate of increase and an exponential rate of increase. The linear assumption was that the rate of clearing from 1981 to 1982 was the same as the rate between 1978 and 1981. On that basis 552 km^2 were cleared between 1981 and 1982 and a total of 3,752 km^2 had been cleared. If the exponential increase continued to be valid, the clearing in 1982 reached 1,120 km^2/yr and the total area cleared was 4,320 km^2. Using these two methods the range of estimates of total area cleared within one LANDSAT scene up to 1982 was 3,752 to 4,320 km^2 or between 12% and 14% of the entire LANDSAT scene.

A comparison of the results of AVHRR and the adjusted LANDSAT data show that the AVHRR data underestimated cleared land by 2 to 18%. Because of the adjustments to the LANDSAT data, however, a calibration with the finer

resolution LANDSAT data is tenuous. The results suggest
that the calibration for this area would be close to 1.0.
We used these data to estimate total deforestation for
the entire state of Rondônia for 1982. The AVHRR data were
used to define the area of cleared forest as those pixels
with digital numbers (DN) greater than one standard
deviation from the mean value for the entire area of
Rondônia. On that basis 11,400 km^2 of forest had been
cleared within the areas of Rondônia that were being
colonized along the central highway, BR-364. This may be
an underestimate because about 5% of southern Rondônia was
cloud-covered on the AVHRR image. Recent maps show that
there are few roads in the cloud-covered region and that a
large portion of the cloud-covered region is within the
Pedras Negras forest reserve (recinded in 1981) and in the
Guaporé Biological reserve (World Bank, 1981).

Other, less detailed appraisals lie in the same general
range. Tucker et al. (1983) have estimated the area of
forest cleared by 1982 along the main axis of immigration
and colonization, highway BR-364, to be between 9,200 and
11,000 km^2. An extrapolation is possible from the
official Brazilian estimates of forest cleared in Rondônia.
Using estimates of forest clearing in 1975 (1,217 km^2
according to Brazil), 1978 (4,185 km^2 according to
Tardin et al, 1980), and 1980 (7,579 km^2 according to
Fearnside, 1984) the exponential rate of increase would
predict 13,517 km^2 of forest clearing by 1982. Finally,
using simple photointerpretation techniques and the NOAA
AVHRR image, we estimated that 12,390 km^2 of forest (plus
or minus 6%) had been cleared by 1982 in Rondônia.

The transitions in the tropics are profound, especially
when they extend to the displacement of those natural
communities of plants and animals that stabilize the
landscape as the habitat for life. While change in the
area of forests is not the only change of importance, it is
fundamental, progressive, often irreversible and, strangely
enough, difficult to measure. The satellite imagery
available now can be used to provide such measurements.
Its use is the first step in gaining sufficient wisdom to
control the changes while there is time to preserve major
segments of this, the richest habitat on earth.

CONCLUSIONS

1. Rates and areas of recent deforestation can be
determined with both LANDSAT and NOAA AVHRR data.

Using a combination of both types of data has several advantages including the high spatial resolution of the LANDSAT data and the high temporal resolution of the AVHRR data. Information gained can be used to lessen current uncertainties about rates and areas of clearing of tropical forests.

2. Using LANDSAT data from three different dates we determined rates of clearing in the area of one LANDSAT scene (32,000 km^2) in the Brazilian Amazon state of Rondônia. Rondônia is the focus of a major Brazilian colonization scheme and consequently is being deforested very rapidly.

3. Rates of forest clearing in the LANDSAT scene doubled over the 1976 to 1981 period from 27,000 ha/yr to 55,000 ha/yr.

4. No non-forest to forest transformation was found in the LANDSAT scene area presumably because all areas cleared are kept in pasture or agriculture and are not allowed to return to forest.

5. Information acquired from the analysis of the LANDSAT data can be scaled up to cover larger areas with data from the lower resolution NOAA AVHRR series of satellites. Additional research is needed here to produce truly quantifiable results.

6. With data from LANDSAT and the AVHRR, we estimate that a minimum of 11,400 km^2 of Rondônia had been cleared by 1982, about 5% of the entire state. There is no other direct evidence with which to compare this estimate but indirect evidence reveals that this estimate is reasonable but perhaps too conservative.

7. These techniques clearly offer a basis for planning for management of land and water as demands for renewable resources soar over the next years when the human population will expand to six billion and beyond.

ACKNOWLEDGMENTS

This research was sponsored by the Carbon Dioxide Research Division, Office of Energy Research, U.S. Dept. of of Energy, under contract number DEACO5-840R21400 with Martin Marietta Energy Systems, Inc.

REFERENCES

Fearnside, P. M., 1982: Deforestation in the Brazilian 40

Amazon: How fast is it occurring? Interciencia, 7(2), 82-88.

Fearnside, P. M., 1984: A Floresta Vai Acabar?, Ciencia Hoje, 2(10), 42-52.

Instituto Brasileiro de Geografia e Estatistica, 1979: Anuario Estatístico do Brasil-1979, Secretaria de Planejamento da Presidência da República, Rio de Janeiro, 853 pp.

Klemas, V. and M Hardisky, 1983: The use of remote sensing in global biosystem studies, Adv. Space Res., 3(9): 115-122.

NASA, 1983: Land-related Global Habitability Science Issues, NASA Technical Memorandum 85841, 112 pp.

Tardin, A. T., D. C. Lee, R. J. R. Santos, O. R. deAssis, M. P. Barbosa, M. Moreira, M. T. Pereira, C. P. Filho, 1980: Subprojeto Desmatamento Convenio IBDF/CNP-INPE (Instituto De Pesquisais Espaciais, São José Dos Campos, Brazil), 44 pp.

Tucker, C. J., B. N. Holben, and T. E. Goff, 1983: Forest Clearing in Rondônia, Brazil as detected by NOAA7 AVHRR Data, NASA Technical Memorandum 85018, 30 pp.

Woodwell, G. M., R. A. Houghton, T. A. Stone, and A. B. Park, in press. Changes in the Area of Forests in Rondônia, Amazon Basin, Measured by Satellite Imagery. In: The Global Carbon Cycle: Analysis of the Natural Cycle and Implications for the Next Century. J. R. Trabalka and D. E. Reichle (eds.) Springer Verlag, New York.

Woodwell, G. M. (ed.), 1984: The Role of Terrestrial Vegetation. In: The Global Carbon Cycle: Measurement by Remote Sensing. Scope 23, Wiley and Sons, Chichester, 247 pp.

Woodwell, G. M., J. E. Hobbie, R. A. Houghton, J. M. Melillo, B. J. Peterson, G. R. Shaver, and T. A. Stone, 1983: Deforestation Measured by LANDSAT: Steps Toward a Method. Technical Report TR005, U. S. Department of Energy, Washington, D. C., 62 pp.

World Bank, 1981: Brazil, Integrated Development of the Northwest Frontier. Latin America and the Caribbean Regional Office, The World Bank, Washington, DC. 101 pp.

Michael B. McElroy, Steven C. Wofsy

4. Tropical Forests: Interactions with the Atmosphere

ABSTRACT

Trace gases such as N_2, NO, CH_4, CO_2, CO, and O_3 play an important direct and indirect role in regulating the climate of the earth. They determine the quantity and quality of solar radiation reaching the planetary surface, influencing to a large extent the chemical environment for life. The composition of the atmosphere reflects the overall metabolism of the biosphere, with a particularly important contribution from the tropics. There is compelling evidence that the composition of the atmosphere is changing. Reasons and possible implications for this change are discussed in the context of current understanding of the atmosphere-biosphere as an interactive system.

INTRODUCTION

The last fifteen years have seen a revolution in atmospheric chemistry. This has come about as a consequence of three parallel developments: studies of stratospheric ozone, studies of tropospheric chemistry, and increased attention directed to possible anthropogenic impacts on climate. We shall review these developments, and show that they have an important feature in common, the recognition that interactions with the biosphere can play a critical role in determining the physical and chemical state of the atmosphere. As we shall see, tropical ecosystems are particularly important in this context.

Tropical systems include a large fraction of the biomass of the earth. Agricultural and industrial

33

development over the past several centuries has involved mainly exploitation of land at middle to higher latitudes of the northern hemisphere. It is clear, though, that the emphasis is switching now to the tropics. It is essential that our understanding of the complex ecosystems which characterize this region advance to keep pace with development.

We shall argue that the atmospheric chemist has a role to play in acquiring the necessary data, that he or she can contribute in an important way to defining the variety of subtle links that bind the living and nonliving parts of the planet. The atmospheric chemist is motivated to do so by questions arising from purely disciplinary considerations. It appears that concentrations of atmospheric carbon dioxide, carbon monoxide, methane, and nitrous oxide are increasing on a global scale at present. We need to understand the processes responsible for this change, and many of the essential clues may reside in the tropics. Atmospheric chemists and terrestrial ecologists have much to learn from each other. Cross disciplinary cooperation is essential if the resources of the planet are to be protected for present and future generations.

CHEMISTRY OF THE STRATOSPHERE

Most of the earth's ozone resides in the stratosphere, at altitudes between roughly 20 and 40 km. The gas is formed by reaction of oxygen atoms with O_2,

$$O + O_2 + M \rightarrow O_3 + M \tag{1}$$

with O atoms derived ultimately from dissociation of O_2,

$$hv + O_2 \rightarrow O + O \tag{2}$$

Reaction (2) is initiated by sunlight at wavelengths below about 240 mm.

It is convenient to identify a family of compounds, odd oxygen, composed, of $O + O_3$. Reaction (2) provides the only significant source for odd oxygen in the stratosphere. It is removed by

$$O + O_3 \rightarrow O_2 + O_2, \tag{3}$$

and by indirect paths such as

$$OH + O_3 \rightarrow HO_2 + O_2 \tag{4a}$$

$$HO_2 + O \rightarrow OH + O_2, \qquad (4b)$$

$$NO + O_3 \rightarrow NO_2 + O_2 \qquad (5a)$$

$$NO_2 + O \rightarrow NO + O_2, \qquad (5b)$$

$$Cl + O_3 \rightarrow ClO + O_2 \qquad (6a)$$

$$ClO + O_3 \rightarrow Cl + O_2 \qquad (6b)$$

and

$$Br + O_3 \rightarrow BrO + O_2 \qquad (7a)$$

$$BrO + O \rightarrow Br + O_2 \qquad (7b)$$

The net effect of reactions sequences (4) - (7) is equivalent to that of the direct reaction (3). The rate for (3) is enhanced accordingly by trace quantities of OH, NO, Cl, and Br. These compounds serve as catalysts for (3).

Ozone, though a minor constituent of the atmosphere, plays an important role in regulating the conditions for life on earth. Solar radiation between 310 and 240 mm is absorbed by O_3, mainly in the stratosphere, shielding the surface from potentially harmful radiation. A large column density of O_3 is biologically beneficial in this sense. High concentrations of O_3 in surface air, however, can have a negative impact on living systems: elevated levels are known to result in marked reductions in the productivity of plants in both natural and agricultural systems (see, for example, Heck et al., 1982).

Recent attention directed to the study of O_3, both in the stratosphere and troposphere, reflect the view that human activities can result in significant changes to the abundance and distribution of O_3. These concerns were raised first in the early 1970's (Crutzen, 1970; Johnston, 1971). The issue then involved exhaust gases of high flying supersonic aircraft, in particular the possibility that additions of NO to the stratosphere could enhance the role of (5) with consequent reduction in the abundance of O_3. Our understanding of stratospheric chemistry was quite rudimentary in 1970. We had yet to recognize the role of NO and Cl in the unperturbed environment. The possibility of man-induced changes to the system served,

though, to stimulate an intensive research program, funded
initially by the U.S. Department of Transportation and
subsequently by NASA. This led to a much clearer
appreciation for the complexity of the natural stratosphere
and its links to the troposphere and biosphere.

We know now that NO plays an important role in the
removal of O_3 under natural conditions. It is derived
mainly from oxidation of N_2O (Crutzen, 1971; Nicolet and
Vergison, 1971; McElroy and McConnell, 1971).

$$O(^1D) + N_2O \rightarrow NO + NO, \qquad (8)$$

where the metastable $O(^1D)$ is produced by photolysis of
O_3,

$$hv + O_3 \rightarrow O(^1D) + O_2. \qquad (9)$$

Nitrous oxide is formed by microbial activity in soils and
aquatic systems, a byproduct of both oxidation
(nitrification) and reduction (dentrification) of fixed
nitrogen (Yoshida and Alexander, 1971; Payne, 1982) with a
contribution also from combustion (Weiss and Craig, 1976;
Pierotti and Rasmussen, 1976). We can calculate the rate
at which N_2O is removed from the atmosphere: the
dominant sink is photolysis in the stratosphere. We may
conclude, thus, with some confidence that the global source
of N_2O amounts to about 10^7 tons N yr^{-1}; the
associated uncertainty is no more than about $\pm$ 30%.
Studies of stratospheric chemistry, in this manner, provide
an important window to the global nitrogen cycle. Related
studies of marine biogeochemistry allow us to argue that
the ocean is not the dominant source of N_2O (Elkins et
al., 1978; Cohen and Gordon, 1979). It follows that
terrestrial systems must dominate. Our task is to home in
on the more significant source regions. As it turns out,
tropical systems are particularly important in this regard.

Questions regarding effects of high-flying aircraft
were followed by concerns that decomposition of industrial
chlorocarbons, notably CF_2Cl_2 and $CFCl_3$, could pose
an even larger problem for stratospheric O_3 (Molina and
Rowland, 1974). These gases are relatively inert and are
used extensively as propellants in aerosol spray cans, as
solvents, and as working fluids in refrigerators. Their
lifetimes in the atmosphere exceed 50 years. They are
removed mainly in the stratosphere, by photolysis, with
consequent release of constituent chlorine. On a mole per
mole basis, chlorine, through reaction (6), is more

effective than NO as a catalyst for removal of O_3. The impact of chlorocarbons on O_3 depends on the subsequent distribution of chlorine between the reactive radical forms Cl and ClO and less reactive species such as HCl and $ClNO_3$. Chlorine radicals can be removed by reaction of Cl and CH_4 forming HCl,

$$Cl + CH_4 \rightarrow HCl + CH_3, \tag{10}$$

or by reaction of ClO with NO_2 forming $ClNO_3$,

$$ClO + NO_2 + M \rightarrow ClNO_3 + M. \tag{11}$$

These sinks, however, are only transitory. Reactive radicals are reconstituted by reactions such as

$$OH + HCl \rightarrow H_2O + Cl, \tag{12}$$

and

$$hv + ClNO_3 \rightarrow ClO + NO_2. \tag{13}$$

Methane has an important influence on the chemistry of the stratosphere over and above its role as a sink for Cl. The stratosphere is very dry: the abundance of H_2O amounts to only a few parts per million of the total air above the tropopause. The factors which determine the level of stratospheric H_2O are as yet poorly understood. The conventional explanation suggests that air enters the stratosphere mainly at low latitudes where convection is most intense and where the tropopause is highest and coldest. Recent observations indicate, however, that the tropical tropopause is not cold enough, at least on average, to account for the exceptionally low values observed for the concentration of stratospheric H_2O. Exceedingly vigorous convection in localized regions of the tropics may provide the answer. It has been suggested that cumulus convection over Indonesia, for example, or the Western Amazon Basin, can penetrate to altitudes as high as 20 km (Kley et al., 1982; Danielsen, 1982 a, b). The tops of associated clouds are extremely cold and offer a potential cold trap to remove H_2O from the lower stratosphere. The abundance of stratosphere H_2O may be linked therefore in a rather direct way to the climate of particular regions of the tropics. Transfer of CH_4 and subsequent oxidation at higher levels of the atmosphere could provide a mechanism whereby hydrogen, and subsequently H_2O, could bypass the tropical cold trap. We would expect an increase in CH_4 to lead to an increase in stratospheric H_2O with consequent effects on stratospheric chemistry. The concentration of OH would respond to changes in H_2O through the reaction

$$O(^1D) + H_2O \rightarrow OH + OH, \qquad (14)$$

the dominant source of stratospheric OH.

Assessment of the effects of human activity on stratospheric O_3 requires information on release rates for catalysts such as Cl and Br. Data are needed to define the changing composition of the background atmosphere, to elucidate factors responsible for trends observed in the concentrations of N_2O, CH_3 and H_2O. Changes in CO_2 may also be important in that an increase in CO_2 will tend to cool the stratosphere, altering rates for several key reactions, offsetting, at least partially, reductions in O_3 caused by elevated levels of NO, Cl, and Br.

CHEMISTRY OF THE TROPOSPHERE

The chemistry of the troposphere seemed relatively simple fifteen years ago. We thought then that photochemistry was confined mainly to cities and other sites of major industrial activity, where complex mixtures of hydrocarbons, carbon monoxide, and oxides of nitrogen could react under the influence of sunlight to produce high concentrations of O_3 and other pollutants. Ozone was believed to form by reactions such as

$$RO_2 + NO -- RO + NO_2 \qquad (15)$$

followed by

$$h\nu + NO_2 -- NO + O, \qquad (16)$$

and (1), where R is a suitable organic radical. It was thought that ozone in the clean environment was derived mainly from the stratosphere, entering the troposphere through gaps in the tropopause at mid-latitudes, to be removed ultimately by heterogeneous reactions at the surface. All of that changed in the early 1970's with the recognition (Levy, 1971) that (9) followed by (14) could provide a significant source of OH, not only in the stratosphere, but also in the troposphere.

Fig. 1.1   View of tropical rain forest near Manaus.

Fig. 1.2   Opening of fruit during Brazil nut harvest.

Fig. 1.3   The fruits of the cacau tree (<u>Theobroma cacau</u>
L.), one of the many Amazonian plants of economic
importance.

Fig. 3.2(a)   INPE LANDSAT scene 176173-125857 from 21 June 1976.   See Fig. 3.1 for location.   Grid patterns are roads and cleared forests within INCRA colonization areas. This scene is about 185 x 185 km (From Woodwell et al., in press).

Fig. 3.2(b)   INPE LANDSAT scene 378216-13336 from 4 August 1978 covering the same area as Fig. 3.2(a). Increased forest clearing is evident when compared to Fig. 3.2(a) (From Woodwell et al., in press).

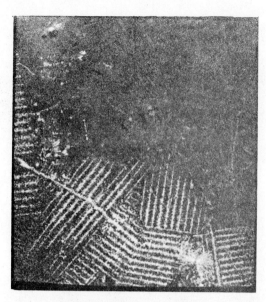

Fig. 3.2(c)   INPE LANDSAT scene 281137-133031 from 17 May 1981 covering the same area as Fig. 3.2(a) and 3.2(b). Extensive new forest clearing is evident by comparison with figs. 3.2(a) and 3.2(b) (From Woodwell et al., in press).

Fig. 3.3 July 1982 NOAA7 AVHRR thermal image. The grids in the center clearly show some of the colonization areas of central Rondônia.   Deforested regions show a warmer area, probably due to canopy loss, lower rates of evapotranspiration, and soil warming.   They contrast strongly with the cooler forested areas (From Woodwell et al., in press).

Fig. 6.2(a) Primary lowland rain forest at Mentoko camp looking Southwest across the Sangata River, photographed in 1979. (b) Exactly the same scene photographed in September 1983, four months after fire.

Fig. 6.2(c) Burned primary forest on dry ridge, showing dead stems of all sizes still standing after fire, and the numerous spider webs joining them (September 1983). (d) Typical condition of crown dieback showing the reduced foliage of many surviving trees.

The chemistry of the troposphere is fueled by $O(^1D)$, present with an abundance of less than 10^{-19} of the total atmosphere, and by OH, representing only about 10^{-13} of the air near the ground. Hydroxyl radicals are removed in the troposphere mainly by reactions with CO and CH_4,

$$OH + CO \rightarrow H + CO_2 \tag{17}$$

and

$$OH + CH_4 \rightarrow H_2O + CH_3. \tag{18}$$

These reactions represent the dominant sinks not only for OH, but also for CO and CH_4. Recent studies indicate concentrations of both CO and CH_4 are increasing in the contemporary atmosphere.

An increase in CO is likely to result in a decrease in OH (Wofsy, 1976), and consequently an increase in the lifetime for CH_4 in the atmosphere. A portion of the rise in CH_4 in the atmosphere may be attributed to this effect (Sze, 1977). The observed increase in CO (Khalil and Rasmussen, 1984) is due at least in part to combustion of fossil fuel: the concentration of the gas in the Northern Hemisphere is almost twice that in the Southern Hemisphere. There is evidence, however, for additional production in the tropics (Logan et al., 1981; Pinto et al., 1983). The tropical source includes a contribution from burning of vegetation associated with land clearance and management of agricultural land. We expect production also from the oxidation of biogenic hydrocarbons such as isoprene. Current estimates suggest that the indirect biospheric contribution could be comparable to the direct source from burning of fossil fuel.

It is possible, from knowledge of OH, to obtain estimates for the global sources of CO and CH_4. The concentration of OH may be obtained in two ways, either from first principles, computing the balance of production and loss, or, more empirically, by studying the behavior of a gas whose source is known and whose loss is controlled by OH. Observations of CH_3CCl_3 provide an opportunity for the second approach, offering a check on results obtained using the first (Lovelock, 1977). To the best of our knowledge there are no natural sources of significance for CH_3CCl_3. The gas is an industrial product, vented to the atmosphere at a rate which is known to better than 30%. A relatively straightforward analysis (Prinn et al., 1983) yields a lifetime of approximately six years, implying an average concentration for OH of about 10^6 cm^{-3}. This suggests that the global removal rate for CH_4 is about 320 tons C yr^{-1}, with an uncertainty of about $\pm$ 30%, and indicates that as much as 1% of the net amount of carbon fixed annually by photosynthesis is oxidized ultimately by processes in the troposphere initiated by reaction of CH_4 with OH (Logan et al., 1981). The issue is more complicated for CO since the concentration of this gas varies appreciably over the globe, as we might expect given its comparatively brief lifetime, a few months at most. We estimate a global source of CO of about 940 tons C yr^{-1}, with an associated uncertainty of about $\pm$40%, using an updated version of the model presented by Logan et al. (1981). The model calculations of the global OH distribution yield results in excellent agreement ($\sim$15%) with the analysis based on the global budget of CH_3CCl_3.

Oxidation of CO, CH_4, and other hydrocarbons can provide an important source of hydrogen-bearing radicals, leading either to production or destruction of O_3 (Logan et al., 1981; Chameides and Walker, 1973; Fishman and Crutzen, 1978). Reaction (17), for example, is followed by

$$H + O_2 + M \rightarrow HO_2 + M. \qquad (19)$$

with HO_2 removed either by

$$HO_2 + O_3 \rightarrow OH + O_2 + O_2. \qquad (20)$$

or by

$$HO_2 + NO \rightarrow OH + NO_2. \qquad (21)$$

Reaction (21) results in production of O_3, since NO_2 is readily dissociated by visible sunlight (Reaction (6)), with O atoms reacting with O_2 via (1) to form O_3. Reaction (20), on the otherhand, represents a sink for O_3. The balance between production and removal of O_3 depends on the abundance of NO relative to O_3. If the concentration of NO is high, above about 70 parts per trillion (ppt), as it is for example in cities and over large regions of the continents, oxidation of CO and hydrocarbons leads to net production of O_3. Otherwise, removal of CO and hydrocarbons, results in consumption of O_3. The abundance and spatial distribution of tropospheric O_3 is linked thus in a rather direct way to the processes which control the levels of atmospheric CO, hydrocarbons, and NO.

Oxides of nitrogen are formed as byproducts of combustion, either at the expense of nitrogen in fuel, or by fixation of atmospheric N_2. The temperature of combustion plays an important rôle in determining whether the combustion process is a net source of fixed N or merely a means for volatilization of preexisting N. Fixation of N_2 occurs at temperatures above about 1500°K, which can arise for example in the combustion chambers of trucks and automobiles, or in power plants. Burning of vegetation, however, can transfer significant quantities of N from biomass to the atmosphere, providing an essential ingredient for smog in otherwise pristine environments. Lightning represents an additional source of NO. There is a potentially large contribution also from microbial oxidation of ammonium in soils (Galbally and Roy, 1978; Lipschultz et al., 1981).

The combination of fire-derived NO and hydrocarbons with sources of natural origin can result in significant production of O_3. There are indications that these conditions are increasingly common in the tropics. Recent studies (Crutzen et al., 1985; Delany et al., 1985) suggest that the concentration of O_3 is suprisingly high in central Brazil during the dry season when burning is most common, large enough, perhaps, to approach phytotoxic levels. This matter merits further attention in light of its potential importance to the continuing health of tropical ecosystems.

CLIMATIC CONSIDERATIONS

There is an excellent, essentially continuous, record for the change in atmospheric CO_2 over the past 25 years. The concentration has risen from about 315 ppm in 1958 to almost 345 ppm today (Keeling et al., 1976). Much of the rise can be attributed to release of CO_2 from combustion of fossil fuel. There are problems, however, in accounting for details of the observed trend. The rise is less than one might anticipate given the magnitude of the fossil source. The difficulty is compounded in that we might have expected that, over the period of observations, burning of vegetation could have provided a source of CO_2 perhaps comparable to that from fossil fuel (Woodwell, 1978; Stuiver, 1978).

Estimates for the source due to burning are based ultimately on data defining the area of land cleared annually for agriculture, mainly in the tropics. There are reasons to believe that at least some of the values published recently for the associated release of CO_2 are too high (Seiler and Crutzen, 1980). The net source of CO_2 depends on the abundance of carbon stored in both the biomass and soils of the area subjected to fire, and on the fraction of bound carbon released to the air. We must account also for uptake of CO_2 in areas recently cleared but subsequently abandoned and allowed to return to forest. There are major gaps in our knowledge of the carbon content of tropical ecosystems and further uncertainties in our ability to predict the fractional yield of CO_2 from fire. A significant portion of the carbon burnt may be converted to charcoal (Seiler and Crutzen, 1980) which is relatively inert and could be stored for long times in soils or in river and estuarine sediments. Careful studies of tropical systems are needed to document conditions before, during, and after fire.

We need to define better the chemical and ecological influences responsible for change. We need also improved data on the extent of the land area burnt annually. Observations from space could play an invaluable role but must be carefully coordinated with ground-based data and analysis if their potential is to be realized. The impact of space observations to date has been relatively disappointing.

An impressive body of information has accumulated recently to suggest that fluctuations in CO_2 may have played an important role in regulating at least some of the major changes in climate of the past. The level of CO_2 was approximately 200 ppm during the last ice age. It rose by about 50% in only a few thousand years, 18,000 years ago, ushering in the present interglacial (Delmas et al., 1980; Neftel et al., 1982). We can reconstruct the history of CO_2 back to almost 60,000 years before present using air trapped in bubbles of ancient ice preserved in Greenland and Antarctica. A more indirect technique, based on analysis of the isotopic composition of carbon in the skeletons of marine organisms in ocean sediments, allows us to extend the record even further, to about 600,000 years before present (Pisias and Shackelton, 1984). The correlation with climate is striking. High CO_2 is associated invariably with warm conditions, low CO_2 with cold. This suggests that the geologic record can provide an important means to check and refine models for the present climate, that it can enhance our ability to predict the response of the atmosphere-biosphere system to contemporary changes in CO_2. We cannot afford to ignore the legacy of the past if we are to chart a wise course to the future.

Carbon dioxide is but one of several gases with the potential to raise the temperature of the earth. Infrared radiation from the planetary surface is absorbed also by CH_4, N_2O, O_3, and by the industrial chlorocarbons, CF_2Cl_2 and $CFCl_3$. On a molecule per molecule basis, these gases are even more efficient than CO_2. According to calculations, indeed, the increased concentrations of these gases over the past decade may have warmed the climate by an amount comparable to CO_2 (Lacis et al., 1981). The complex nature and diverse reasons for the changes observed in the composition of the atmosphere require an analysis more comprehensive than any performed to date. We must begin to think of the earth as an interactive system. Changes in the atmosphere can induce changes in the ocean and biosphere, and in the storage and transfer of H_2O, which may affect the function of all of the important compartments of the planet. Tropical ecosystems are particularly sensitive and especially important, we believe, in determining the overall global response to the diverse range of disturbances introduced by man.

SOURCES AND SINKS OF MAJOR TRACE GASES

As noted earlier, nitrous oxide is removed from the atmosphere mainly by photolysis in the stratosphere and by reaction with $O(^1D)$, reactions (8) and

$$N_2O + hv \rightarrow N_2 + O. \qquad (22)$$

The rate for destruction is calculated to be 10.5 ($\pm$3) x 10^6 tons N yr^{-1}, using observed distributions for N_2O and calculated rate coefficients for (22) and (8). The rate of increase in the atmosphere is currently 0.7 ± 0.1 ppb yr^{-1} (3.5×10^6 tons N yr^{-1}), with an indication that growth may have accelerated since 1965 (Weiss, 1982; Weiss and Craig, 1976). Total emissions for N_2O amount therefore to 14 ± 3 x 10^6 tons N yr^{-1}, as shown in Fig. 4.1. The magnitude of the annual increase, though small, implies a discrepancy in excess of 30% between current sources and sinks, a consequence of the long atmospheric residence time (150 yr) for N_2.

Nitrous oxide is an obligatory free intermediate in dentrification (Payne, 1982).

organic matter + NO_3^- → NO_2^- → NO
→ N_2O → N_2 (23)

Sequential reduction of the nitrogen atom in (23) provides a respiratory path for a wide variety of bacteria under anaerobic conditions. Denitrification is most often observed in environments isolated from atmospheric oxygen and supplied with abundant sources of oxidizable dentrital material, organic-rich sediments, flooded soils, and closed ocean basins, for example. Such systems were thought at one time to be the principal sources for atmospheric N_2O, but this idea turned out to be incorrect. Anaerobic ecosystems contain typically very low concentrations of N_2O, indicating that virtually all the N_2O produced in (23) is consumed in situ (Cohen and Gordon, 1978; Hashimoto et al., 1983).

Significant quantities of N_2O are produced, however, by a variety of aerobic environments. The most intense emissions are associated with rapid oxidation of organic matter, and it appears that the N_2O is produced mainly as a by-product of primary nitrification,

$$NH_4^+ + \frac{3}{2}O_2 → H_2O + NO_2^- + 2H^+$$

(24)

This process, carried out by a small group of autotrophic bacteria, yields 1-3 molecules of N_2 per 1000 nitrite molecules under fully aerobic conditions. The yield of N_2O increases dramatically under low-oxygen conditions, rising to 10% of the nitrite production rate for partial pressures of O_2 below 0.01 atm (Goreau et al., 1980). It is interesting to note that nitrifying bacteria also produce NO, with release rates similar to N_2O (Yoshida and Alexander, 1971; Lipschultz et al., 1981). We expect, therefore, that soils producing large quantities of N_2O might also form copious amounts of NO, although direct observations of NO release are presently scant (Galbally and Roy, 1978).

Soils in tropical forests emit N_2O at rates far in excess of those observed in most other environments, as shown by recent measurements from forest sites in Brazil, Ecuador, Puerto Rico, and New Hampshire (Goreau, 1981; Keller et al., 1983; Keller et al., to be published). Release rates for tropical forests average about 2×10^{10} molecules N_2O cm^{-2} sec^{-1}, as compared to $1-2 \times 10^9$ in New Hampshire. An interesting inverse relationship was observed between fluxes of CH_4 and emissions of N_2O. Sites which consumed atmospheric CH_4 had the highest emission rates for N_2O, while sites which emitted CH_4 had lower emissions of N_2O. The latter tended to be waterlogged, or nearly so. Microbial production of CH_4 is a strictly anaerobic process, while consumption generally requires oxygen. The observed anticorrelations of N_2O and CH_4 emissions provide support for the view that N_2 is produced in forest soils at least in part by an oxidative process such as (2), although a variety of other microbial processes might contribute (Yoshida and Alexander, 1970). The large emissions observed for N_2O may be interpreted to indicate rapid oxidation of mineral nitrogen in tropical forest soils, suggesting that the nitrogen cycle in these systems is less conservative than commonly assumed.

Concerns arose during the 1970's (McElroy et al., 1977) that use of nitrogenous fertilizer would artifically enhance biogenic emissions of N_2O, leading perhaps to increased concentrations in the atmosphere. Recent investigations indicate that fertilization with NH_4^+ or urea does indeed stimulate emission of N_2O, although the yield is relatively small (Hutchinson and Mosier, 1976; McKenney et al., 1978; Bremmer et al., 1980). Between 0.1 and 0.5% of the reduced nitrogen in fertilizer is converted to N_2O within a few weeks of application, with highest efficiency for conversion at highest rates of fertilization. The ultimate release of N_2O after fertilization could be higher, however, since fixed nitrogen is likely to be assimilated into organic material and re-oxidized a number of times before it is lost eventually from the soil.

Combustion introduces another important anthropogenic source for N_2O, (Pierotti and Rasmussen, 1976; Weiss and Craig, 1976). Recent studies (Perry, 1984) have shown that N_2O in flames is produced from fuel nitrogen by the rapid reaction,

$$NO + NCO \rightarrow N_2O + CO. \qquad (25)$$

Data from this laboratory (W.M. Hao, to be published) indicate that approximately 10% of the nitrogen in fuel is converted to N_2O during a typical combustion process. Somewhat smaller amounts of N_2O are produced in fuel-rich flames, in two-stage combustors for example or in smouldering fires, reflecting, apparently, subsequent reactions of N_2O with reduced species such as H.

Figure 4.1 sumarizes current understanding of sources for atmospheric N_2O. Estimates for the marine source are based on the nitrification process (Elkins et al., 1978; Cohen and Gordon, 1979), using observations of accumulation of N_2O and depletion of O_2 in marine waters, supported by extensive data for N_2O dissolved in surface waters of the world's oceans (R. Weiss, 1983). The value for combustion is derived from data defining the composition and utilization of various fuels, assuming that 10% of fuel nitrogen is converted to N_2O. The entry for fertilized agricultural lands assumes application of chemical fertilizer at the rate of 40×10^2 tons N yr^{-1} and a similar rate for use of manures, with an overall yield for N_2O of $1+0.5$%. The value for forests reflects an average of several hundred direct measurements of forest soils in New Hampshire, Puerto Rico, Ecuador, and Brazil (Keller et al., 1983; Keller et al., to be published).

It appears that anthropogenic processes account for about one third of current emissions. The figure implies a pre-industrial concentration of N_2O approximately 10-20% lower than today. If the present pattern of emissions persists, the abundance of atmospheric N_2O should grow slowly to about 400 ppb. However, there is little reason to project that emissions should remain constant in the future. Sources associated with combustion and with intensive agriculture are likely to increase, and we might expect increased fluxes of N_2O from tropical forests disturbed by exploitation. On the other hand, the source from crop land and pasture may be smaller than from undisturbed systems. A much improved understanding is needed to predict future emissions of N_2O and studies of tropical forests are clearly important to this aim.

Reaction of OH with CH_4 removes about 320×10^6 tons C of methane per year (see discussion above). Biological uptake of CH_4 has been observed for a wide variety of aerobic soils, but the rate for global consumption ($0+5 \times 10^6$ tons C yr^{-1}) is relatively small (Keller et al., 1983; Seiler et al., 1984). About 45×10^6 tons C accumulate annually in the atmosphere, accounting for a total atmospheric throughput of CH_4 equal to $375+105 \times 10^6$ tons C yr^{-1}, as shown in Fig. 4.2.

Sources for atmospheric CH_4 are difficult to quantify. Methane is produced by fermentative bacteria requiring a strictly anaerobic environment. Production occurs mainly in carbon-rich systems isolated from the atmosphere, for example, flooded forest soils, tundra, bogs and lake bottoms, (Harriss et al., 1981; Martens and

A. <u>Atmospheric</u> <u>burden</u> (10^6 tons as N)   1500

B. <u>Sinks</u> + <u>accumulation</u> (20^6 tons N yr^{-1})
stratospheric photolysis +$_{-1}$reaction with O($_D$)   10.5±3
accumulation (0.7 ppb yr^{-1})   3.5±0.5

                    total   14.0+3.5

C. <u>Sources</u> (10^6 tons N yr^{-1})
oceans   2±1
combustions: coal + oil, 4±1
               biomass, 0.7+0.2

fertilized agricultural lands   4.7+1.2
grasslands   8.1
boreal and temperate forests   0.1
tropical and subtropical forests and woodlands
   (extrapolation of soil data)   7.4(±4)

                    total   15.3(±6.7)

D. <u>Tropical</u> <u>Contricution</u>
burning   100
forest clearing   160
oxidation of hydrocarbons   150

                    total   410

Fig. 4.1   Nitrous oxide (1984 concentration 303 ppb). (From Logan et al., 1981; updated by Logan et al., 1984.)

A. <u>Atmospheric burden</u> (3500×10 tons as C yr^{-1}

Reaction with OH	320 ± 100
Uptake by dry soils	40 ± 25
Accumulation (20 ppb/yr)	45 ± 10
total	405 ± 135

B. <u>Sources</u> (10^6 tons C yr^{-1})

ocean	13
tundra	12
biomass burning	25
natural gas loss + coal mining	40
rice paddies	95
cattle	120
termites	50
wetlands + minor sources (by difference)	50(+?)
total	405

C. <u>Tropical Contribution</u> (10^6 tons N yr^{-1})

biomass	0.6
soil emissions	7.4
total	8.0

Fig. 4.2 Methane (1984 concentration 1630 ppb). (Adapted from Khalil and Rasmussen, 1983; Ehhult et al., 1983; and Logan et al. 1981; updated by Logan, 1984).

Val Klump, 1980). Methanogenic bacteria flourish also in the digestive tracts of ruminants and termites, where they comprise an essential part of the intestinal flora (Hutchinson, 1954; Zimmerman et al., 1982).

Estimates of sources for atmospheric methane, based mostly on Khalil and Rasmussen (1983), are summarized in fig. 4.2. Large scale field measurements are lacking for almost all of the individual sources listed here; results are correspondingly uncertain. Agricultural activities contribute 50% of the total emission, with an additional 20% from fossil fuels and from burning of biomass. It appears that the tropics account for about one fourth of the current global source, two thirds of the unperturbed background.

Analyses of air trapped in polar ice cores indicate that the concentration of atmospheric methane was constant for many thousands of years prior to 1600 AD, and that it began to rise rapidly about 250-400 years ago (Craig and Chow, 1982; Rasmussen et al., 1982). Methane appears to have increased by about a factor of 2.2 since the industrial revolution. The increase may be compared with our estimate for current sources, more than 70% of which are attributed to human activity. As noted earlier, increasing emissions of CH_4 and CO tend to suppress the level of OH (Sze, 1977). The concentration of CH_4 may be expected to rise by more than a factor of 4 in response to the source scenarios shown in figs. 4.2 and 4.3, all other factors being equal. The inconsistency could be resolved if it is assumed that natural sources decline while anthropogenic emissions increase, or, otherwise, if anthropogenic emissions of NO_x result in an increase in the global concentration of OH. These alternative views have quite different implications for models of atmospheric chemistry.

Direct measurements of methane emission from major ecosystems could help resolve the many unanswered questions about the cycle of atmospheric CH_4. It is difficult, unfortunately, to make measurements which represent all the methane released from complex ecosystems. A large fraction of the methane produced in anaerobic environments may be oxidized at the aerobic/anaerobic interface. Emissions to the atmosphere may proceed largely by indirect or sporadic routes, by ebullition of bubbles, for example, or through the stems of vascular plants rooted in the anoxic zone (Cicerone and Shetter, 1981; Martens and Val Klump, 1980).

The magnitude of emissions from tropical forests can be

A.	Atmospheric burden (10^6 tons as C)	200

B. Sinks + accumulation (10^6 tons as C year^{-1})

reaction with OH	820+300
soil uptake	110
accumulation (5.5% yr^{-1})	10
total	940+330

C. Sources (10^6 tons as C yr^{-1})

fossil fuel combustion	190
oxidation of anthropogenic hydrocarbons	40
wood used as fuel	20
oceans	20
oxidation of CH_4	260
forest wild fires (temperate zone)	10
agricultural burning (temperate zone)	10
oxidation of natural hydrocarbons (temperate zone)	100
burning of savanna and agricultural land (tropics)	100
forest clearing (tropics)	100
oxidation of natural hydrocarbons (tropics)	150
total	1060(+?)

D. Tropical Contribution (10^6 tons C yr^{-1})

tropical wetlands, forests and savannas	80
tropical biomass burning	20
total	100

Fig. 4.3 Carbon monoxide (1984 concentrations 30-200 ppb). (Adapted from Khalil and Rasmussen, 1983; Ehhalt et al., 1983; and Logan et al., 1981; and updated by Logan, 1984.)

estimated using our preliminary data (Wofsy, Rasmussen, et al., to be published) for concentrations of N_2O and CH_4 in the atmosphere, together with flux measurements for N_2O. Data were taken in the atmospheric boundary layer of the central Amazon Basin, and samples were also acquired upwind as a coastal site, before the air entered the boundary layer over the tropical forest. Atmospheric mixing processes are linear with respect to trace gas concentrations, and therefore

$$\Phi_{CH_4} = \Phi_{N_2O} \, C_{CH_4} / C_{N_2O}, \qquad (26)$$

where Φ_i denotes the areal mean flux of the ith species $(cm^{-2}sec^{-1})$ and C_i is the excess concentration observed in the Amazon. Here we assume that the sources of CH_4 and N_2O are located sufficiently close to one another that the ratio C_{CH_4} / C_{N_2O}

in the lowest layers of the atmosphere (0-100m) may be regarded as approximately uniform.

Figure 4.1 summarizes preliminary results from this analysis. Atmospheric concentrations were monitored in forest clearings or on large rivers, and at a number of upwind coastal sites. Fluxes of N_2O and CH_4 were measured at distributed sites on aerated forest soils in the tropical forests of Brazil. These soils are believed to be the only major sources of N_2O. They consumed modest amounts of CH_4, as expected. The ecosystems as a whole were found to provide a strong source of CH_4, most likely from wetlands and flood plains which are widely distributed through the forest and perhaps from termites. The mean emission rate for methane was calculated from (26) as 4.3×10^{11} $cm^{-2}sec^{-1}$, suggesting an integrated source of 83×10^6 tons C yr^{-1} for the world's tropical and subtropical lands. Fortuitously, perhaps this result is close to the sum of tropical sources (140×10^6 tons yr^{-1}) given in fig. 4.2.

A similar analysis for other alkanes, such as C_2H_6 and C_3H_8, shows measurable excess concentrations in the Amazon Basin, but emission rates observed for these gases are insignificant on a global scale (see Table 4.1). The higher alkanes appear to originate largely from combustion and from industrial sources, in contrast to CH_4, which is largely biogenic in origin.

The global budget for CO is presented in fig. 4.3, based on estimates by Logan et al. (1981) and Logan (1982), and on updated model calculations by the same authors. The

TABLE 4.1
Estimated mean emission rates for the tropical forest ecosystem from air concentration data

	mean flux from soils ($cm^{-2} sec^{-1}$)	ΔC Manaus-coast (ppb) source a	mean flux from eq. (26) ($cm^{-2} sec^{-1}$)	emissions from source b $31 \times 10^6 km^2$ (tons N or C yr^{-1})	fraction of global source c (%)	fraction of non anthropogenic source d (%)
N_2O	1.6×10^{10}	3.0	-	7.3×10^6	58	90
CH_4	-2.0×10^{10}	80	4.3×10^{11}	83×10^6	20	66
C_2H_6	-	0.5	1.6×10^9	0.6×10^6	2	
C_3H_9	-	0.3	9.2×10^8	0.5×10^6	0.5	

a Based on 10 months of data from this laboratory and from Oregon Graduate Center.

b Area of tropical and subtropical moist forests from Leith (1975).

c Using sinks + total accumulation, figs. 4.1, 4.2, and 4.3

d Using source partitioning from figs. 4.1, 4.2, and 4.3, in conjunction with total current sources from note (c).

model incorporates a careful reconstruction of observed seasonal and latitudinal variations for CO, a requirement arising from the short atmospheric lifetime. The dominant sink for CO is reaction with the hydroxyl radical, while sources involve combustion and photooxidation of hydrocarbons. Direct biogenic emissions are relatively insignificant, in contrast to the case for N_2O and CH_4.

Several sources of CO can be specified with reasonable accuracy, including the contributions from fossil fuel combustion and from oxidation of methane. These account for nearly 50% of the global budget, and since methane is itself largely anthropogenic (fig. 4.2), there can be no doubt that human activities play a dominant role in the cycle of atmospheric CO. The breakdown for other source processes is uncertain, however, although human activities appear to be important here also.

Recent data (Crutzen et al., 1985; Delany et al., 1985) indicate significant pollution of the atmosphere over Brazil during the dry season, attributed to emissions from large-scale agricultural burning. Concentrations of CO as large as 200 ppb were observed over the rain forest, with levels in excess to 400 ppb recorded over the seasonally dry cerrado. These values exceed concentrations found at the same latitude over the oceans by factors of 2-5, and elevated concentrations extend well into the middle troposphere. A similar situation is observed over parts of Africa (Seiler and Junge, 1970) and India (Newell et al., 1981).

The massive emissions of CO due to burning are associated with large inputs of NO_x and reactive hydrocarbons (Crutzen et al., 1984) providing an ideal mix for generations of photochemical smog. Levels of ozone are elevated (Delany et al., 1985), with concentrations between 60 and 80 ppb observed commonly over the cerrado. The abundance of ozone declined just above the forest canopy, indicating rapid reaction with vegetation or with natural olefins emanating from the forest. It is possible that tropical ecosystems may be damaged by deposition of ozone and other associated phytotoxic components of air pollution. Widespread damage from pollution is observed in the forests of Europe and in parts of the United States. It is difficult, however, to identify the principal phytotoxic agents and essentially impossible to anticipate the spread of major tree mortality. While there is as yet no evidence for damage in the tropics, the presence of significant photochemical pollution suggests a need for caution and for careful monitoring of tropical ecosystems, particularly in areas which may be influenced by regional burning.

CONCLUSION

The composition of the atmosphere reflects, as we have seen, the state and metabolism of the global biosphere. The atmosphere is not static. The mix of gases is ever shifting, influenced by a complex array of natural and anthropogenic forces.

Change is most evident for the long lived gases, CO_2, CH_4, and N_2O, and for the various halocarbons, and for CO. Variations in atmospheric composition may be expected to modify climate, the distribution of temperature, rainfall and evaporation over the earth, and to alter the patterns of weather systems and ocean currents. They may result also in pertubations to the composition of short lived reactive gases in the air near the ground, O_3 for example, with potential impact on biospheric productivity. The chemistry of the precipitation may be altered. Even the quantity and quality of solar radiation reaching the earth's surface is subject to change through chemical effects on the stratosphere.

We have learned a lot over the past few years concerning the influence of the biosphere on the atmosphere. Fluxes from the biosphere to the atmosphere can be determined, to better than a factor of 2 in many cases, using techniques of meteorology and atmospheric chemistry. It is more difficult to define the dynamics of the coupled atmosphere-biosphere system, to assess the consequence of changes in the atmosphere for the biosphere and vice versa. This will require a deeper appreciation for the complex interplay of internal and external factors which regulate the structure, dynamics, productivity, and metabolism of major ecosystems.

The tropics are especially important in this connection. Tropical forests account for a relatively large fraction of the world's biomass and an even larger portion of the global productivity. Cycling rates for carbon and nitrogen are rapid, accounting, we believe, for the importance of tropical and subtropical forests in the contemporary budgets of atmospheric N_2O, CH_4, and CO. Tropical systems are the sites of most intense development at the current epoch. They have been studied relatively little in the past. This must change if we are to avoid undesirable, potentially widespread, consequences of human activity in an area which has been up to now comparatively pristine.

The scientific community is ready and eager for a multidisciplinary and coordinated approach to improve our understanding of the earth as a coupled system. Initial discussions of possible initiatives have taken place over the past several years in the United States within NASA (Goody, 1982; McElroy, 1983) and the National Academy of Sciences (N.A.S., 1985, to be published). An international program is under study by the International Council of Scientific Unions (ICSU, 1984). It is clear that the tropics will attract considerable emphasis if these endeavors are to proceed from discussion to action. The problems merit attention and the time to proceed is now.

ACKNOWLEDGMENTS

We are indebted to M. Keller, W. M. Hao, J. A. Logan and M. J. Prather for useful discussions, and R. Rasmussen for results of recent work before publication. This work was supported by NSF grant ATM-81-17009 and NASA grants NSG1=55 and NAGW-359 to Harvard University.

REFERENCES

Bremmer, J. M., S. G. Robbins and A. M. Blackmer, 1980: Seasonal variability in emission of nitrous oxide in soil. Geophys. Res. Lett. 7, 611-643.

Chameides, W. and J. C. G. Walker, 1973: A photochemical theory of tropospheric ozone. J. Geophys. Res. 78, 8751-8760.

Cicerone, R. J. and J. D. Shetter, 1981: Sources of atmospheric methane: measurements in rice paddies and a discussion. J. Geophys. Res. 86, 7203-7209.

Cohen, Y., and L. I. Gordon, 1978: Nitrous oxide in the oxygen minimum of the Eastern Tropical North Pacific: evidence for its consumption during dentrification and possible mechanisms for its production. Deep Sea Res. 6, 509-525.

Cohen, Y., and L. I. Gordon, 1979: Nitrous oxide production in the ocean. J. Geophys. Res. 84, 347-353.

Craig, H., and C. C. Chow, 1982: Methane: the record in polar ice cores. Geophys. Res. Lett. 9: 477-481.

Crutzen, P. J., 1970: The influence of nitrogen oxides on

the atmospheric ozone content. Quant. J. Roy. Met. Soc. 96, 320-325.

Crutzen, P. J., 1971. Ozone production rates in an oxygen-nitrogen atmosphere. J. Geophys. Res. 76, 7311-7320.

Crutzen, P. J., A. C. Delany, J. Greenberg, P. Haagenson, L. Heidt, R. Lueb, W. Pollock, W. Seiler, A. Wartburg and P. Zimmerman, 1985: Tropospheric chemical composition measurements in Brazil during the dry season. J. Atmos. Chem. 2, 233-256.

Danielsen, E. F., 1982a. Statistics of cold cumulonimbus anvils based on enhanced infrared photographs. Geophys. Res. Letter. 9, 601-604.

Danielsen, E. F., 1982b: A dehydration mechanism for the stratosphere. Geophys. Res. Letter. 9, 605-608.

Delany, A. C., P. J. Crutzen, P. Haagensen, S. Walters and A. F. Wartburg, 1985: Photochemically produced ozone in the emission from large scale tropical vegetation fires. J. Geophys. Res., 90, 2425-2429.

Delmas, R. J., J. M. Ascencio and M. Legrand, 1980. Polar ice evidence that atmospheric CO_2 129,000 BP was 50% of the present. Nature 282, 155-157.

Ehhalt, D. H., R. J. Zander and R. A. Lamontagne, 1983: On the temporal increase of tropospheric CH_4. J. Geophys. Res. 88, 8442-8446.

Elkins, J. W., S. C. Wofsy, M. B. McElroy, C. E. Kolb and W. Kaplan, 1978: Aquatic sources and sinks for nitrous oxide. Nature 275, 602-606.

Fishman, J. and P. J. Crutzen, 1978: The origin of tropospheric ozone. Nature 274: 855-857.

Galbally, I. F. and C. R. Roy, 1978: Loss of fixed nitrogen from soils by nitric oxide exhalation. Nature 275, 734-735.

Goody, R., 1982: Global Change: Impacts on Habitability. A report by the Executive Committee of a workshop held at Woods Hole, Mass., June 21-26, 1982. National Aeronautics and Space Administration.

Goreau, T. J., 1981: Biogeochemistry of nitrous oxide. Thesis, Harvard University, 145 pp.

Goreau, T. J., W. A. Kaplan, S. C. Wofsy, M. B. McElroy, F. W. Valois and S.W. Watson, 1980. Production of NO_2^- and N_2O by nitrifying bacteria at reduced concentrations of oxygen. Appl. Environ. Microbiology 40, 526-532.

Harriss, R. C., D. I. Sebecher and F. P. Day, 1981: Methane flux in the great dismal swamp. Nature 247, 673-674.

Hashimoto, L. K., W. A. Kaplan, S. C. Wofsy and M. B.

McElroy, 1983: Transformations of fixed nitrogen in the Cariaco Trench. Deep Sea Res. 30, 575-590.

Heck, W. W., O. C. Taylor, R. Adams, G. Bingham, J. Miller, E. Preston and L. Weinstein, 1982: Assessment of crop loss from ozone. J. Air Poll. Control Assoc. 32, 353-361.

Hutchinson, G. L. and A. R. Mosier, 1976: Nitrous oxide emissions from an irrigated cornfield. Science 205, 1125-1127.

Hutchinson, G. E., 1954: The biogeochemistry of the terrestrial atomsphere. Pages 371-433. In: G. P. Kuiper (ed.). The Solar System II. The Earth as a planet. Univ. of Chicago Press.

International Council of Scientific Unions, Proceedings of the General Assembly, Ottawa, Canada, September, 1984.

Johnston, H. S., 1971: Reduction of stratospheric ozone by nitrogen oxide catalysts from SST exhaust. Science 173, 517-522.

Keeling, C. D., J. A. Adams, Jr., and P. R. Guenther, 1976: Atmospheric carbon dioxide variations at the South Pole. Tellus 28, 552-564.

Keller, M., T. J. Goreau, S. C. Wofsy, W. A. Kaplan and M. B. McElroy, 1983: Production of nitrous oxide and consumption of methane by forest soils. Geophys. Res. Lett. 10, 1156-1159.

Khalil, M. A. K. and R. A. Rasmussen, 1983: Sources, sinks and seasonal cycles of atmospheric methane. J. Geophys. Res. 88, 5131-5144.

Khalil, M. A. K. and R. A. Rasmussen, 1984: Carbon monoxide in earth's atmosphere. Science 244, 54-56.

Kley, D., A. L. Schmettekopf, K. Kelly, R. H. Winkler, T. L. Thompson, and M. MacFarland, 1982: Transport of water through the tropical tropopause. Geophys. Res. Lett. 9, 617-620.

Lacis, A., J. Hansen, P. Lee, T. Mitchell and S. Lebedeff, 1981: Greenhouse effect of trace gases, 1970-1980. Geophys. Res. Lett. 8, 1035-1038.

Leith, H., 1975: Primary Production of the Major Vegetation Units of the World. Ecological Studies 14, (Springer-Verlag, New York, Heidelberg, Berlin).

Levy, H., II, 1971: Normal atmosphere: large concentrations of formaldehyde and radicals predicted. Science. 173, 141-143.

Lipschultz, F., O. C. Zaferiou, S. C. Wofsy, M. B. McElroy, F. W. Valois, and S. W. Watson, 1981: Production of NO and N_2O by soil nitrifying bacteria. Nature 294, 641-644.

Logan, J. A., 1982: Nitrogen oxides in the troposphere: Global and regional budgets. J. Geophys. Res. 88, 10785-10807.

Logan, J. A., M. J. Prather, S. C. Wofsy and M. B. McElroy, 1981: Tropospheric chemistry: a global perspective. J. Geophys. Res. 86, 7210-7254.

Logan, J. A., M. J. Prather, S. C. Wofsy and M. B. McElroy, 1984: pers. communication.

Lovelock, J. E., 1977: Methyl chloroform in the troposphere as an indicator of OH radical abundance. Nature 267, 32.

Martens, C. S. and J. Val Klump, 1980: Biogeochemical cycling in an organic-rich coastal basin. I. Methane sediment-water exchange processes. Geochimica et Cosmochimica Acta 44, 471-490.

McConnell, J. C., M. B. McElroy and S. C. Wofsy, 1971: Natural sources of CO. Nature 233, 187-188.

McElroy, M. B. and J. G. McConnell, 1971. Nitrous oxide: a natural source of stratospheric NO. J. Atmos. Sci. 28, 1095-1098.

McElroy, M. B., S. C. Wofsy, and Y. L. Yung, 1977: The nitrogen cycle: perturbations due to man and their impact on atmospheric N_2 and O_3. Phil. Trans. Roy. Soc. B277, 159-181.

McElroy, M. B., 1983: Global Change: A Biogeochemical Perspective. National Aeronautics and Space Administration.

McKenney, D. J., D. L. Wade, and W. I. Findlay, 1978: Rates of N_2O evolution from N-fertilized soil. Geophys. Res. Lett. 5, 777-780.

Molina, M. and F. S. Rowland, 1974: Stratospheric sinks for cholorfluoromethanes: chlorine atom catalyzed destruction of ozone. Nature 249, 810-812.

National Academy of Sciences, Report of the Committee for an International Geosphere Biosphere Program, to be published, 1985.

Neftel, A., H. Oeschger, J. S. Schwander, B. Stauffer and R. Zumbrunn, 1982: Ice core sample measurements give atmospheric CO_2 contents during the past 40,000 years. Nature 295, 220-223.

Newell, R. E., E. P. Condon and H. G. Reichle, Jr., 1981: Measurements of CO and CH_4 in the troposphere over Saudi Arabia, India and the Arabian Sea during the 1979 International Summer Monsoon Experiment (MONEX). J. Goephys. Res. 86, 9833-9838.

Nicolet, M. and A. Vergison, 1971. L'oxyde azoteux dans la stratosphere. Aeronomica Acta 90, 1-26.

Payne, W. J., 1982: Dentrification, Wiley, New York, 214 pp.

Perry, R., 1984: Paper delivered at 20th Photochemistry Conference, Cambridge, MA., August, 1984.

Pierotti, D. and R. A. Rasmussen, 1976: Combustion as a source of nitrous oxide in the atmosphere. Geophys. Res. Lett. 3, 265-267.

Pinto, J. P., Y. L. Yung, D. Rind, G. L. Russell, J. A. Lerner, J. E. Hansen and S. Hameed, 1983: A general circulation model study of atmospheric methane. J. Geophys. Res. 88, 3691-3702.

Pisias, N. G. and N. J. Shackleton, 1984: Modelling the global climate response to orbital forcing and atmospheric carbon dioxide changes. Nature 310, 757-759.

Prinn, R. G. R. A. Rasmussen, P. G. Simmonds, F. N. Alyea, D. M. Cunnold, B. C. Lane, C. A. Cardelino and A. J. Crawford, 1983: The Atmospheric Lifetime Experiment. 5. Results for CH_3CCl_3 based on three years of data. J. Geophys. Res. 88, 8415-8426.

Rasmussen, R. A., M. A. K. Khalil and S. D. Hoyt, 1982: Methane and carbon monoxide in snow. J. Air Pollution Control Assoc. 32, 176-177.

Seiler, W., R. Conrad and D. Schaffe, 1984: Field studies of the CH_4 emissions from termite nests into the atmosphere and measurements of the CH_4 uptake by tropical soils. J. Atmos. Chem. (1, 171-187.

Seiler, W. and P. J. Crutzen, 1980: Estimates of gross and net fluxes between the biosphere and the atmosphere from biomass burning. Climatic Change 2, 207-248.

Seiler, W. and C. E. Junge, 1970: Carbon monoxide in the atmosphere. J. Geophys. Res. 75, 2217-2226.

Stuiver, M., 1978: Atmospheric carbon dioxide and carbon reservoir changes. Science 199, 253-258.

Sze, N. D., 1977: Anthropogenic CO emissions: Implications for atmospheric $CO-OH-CH_4$ cycle. Science 195, 673-675.

Weiss, R. F., 1982: The temporal and spatial distribution of nitrous oxide. J. Geophys. Res. 86, 7185-7195.

Weiss, R. F., 1983: Personal communication.

Weiss, R. F. and H. Craig, 1976. Production of atmospheric nitrous oxide by combustion. Geophys. Res. Lett. 3, 751-756.

Wofsy, S. C., 1976: Interactions of CH_4 and CO in the earth atmosphere. Annual Review of Earth and Planetary Science 4, 441-469.

Woodwell, G. A., 1978: The carbon dioxide question.

Scientific American 238, 34-43.

Yoshida, T. and M. Alexander, 1970: Nitrous oxide formation by Nitrosomonas Europaea and heterotrophic organisms. Soil Sci. Soc. Am. Proc. 34, 880-882.

Yoshida, T. and M. Alexander, 1971: Hydroxylamine oxidation by Nitrosomonas Europaea. Soil. Sci. 3, 307-312.

Zimmerman, P. R., J. P. Greenberg, S. O. Wandiga and P. J. Crutzen, 1982: Termites: a potentially large source of atmospheric methane, carbon dioxide and molecular hydrogen. Science 218, 563-565.

Eneas Salati, Peter B. Vose,
Thomas E. Lovejoy

5. Amazon Rainfall, Potential Effects of Deforestation, and Plans for Future Research

HYDROLOGY, PRECIPITATION RECYCLING, AND TROPICAL FORESTS

Hydrologists have been inclined to discount the possibilities of land use changes affecting rainfall (Pereira, 1973), as well as to minimize the role of hydrological recycling (UNESCO, 1971). Indeed Linsley (1951) suggested evaporation from the land is not important for continental precipitation, and that deforestation and increased drainage would have little effect on regional precipitation regimes.

There are a number of reasons the potential role of precipitation recycling by tropical forests has been discounted: (1) prior to the great increase in tropical deforestation rates of the last twenty years, many of the areas converted were small relative to the whole; (2) changes in land use have rarely taken place after definitive hydrological and meterological studies, so there is no good "before" and "after" data; (3) regional and subregional vegetation formations have been regarded as defined by soil type, temperature, and the rainfall regime; (4) there has been a tendency to consider evaporation to be about the same for different vegetation types and of significance only when rainfall is a limiting factor; (5) adequate account has not been taken of increased precipitation run off from deforestation and consequent loss of water from the system; (6) in many areas (e.g. southeast Asia and Oceania) the major contribution to precipitation from oceanic water has masked the contribution from recycling.

Studies of the relation of water to tropical forest in the Amazon Basin benefit from the clearly defined region, located on the Equator with a unidirectional source of oceanic water. The basin is horseshoe shaped, with a large plain bounded on the west by the Andes with approximately 4000 m altitude. To the north the basin is limited by the Guianan plateau (with altitudes as high as 1000 m), and to the south by the Central plateau of Brazil (with an altitude about 700 m). Prevailing tradewinds blow almost continuously from the East, penetrating the whole region and bringing oceanic water from the Atlantic. This is only one source of advective water.

Cochrane and Jones (1981) conclude that the savannas of tropical South America occupy a region delimited by climatic potential for growth. This is primarily a wet season potential evapotranspiration regime of about 900 mm yr^{-1} for savanna with a six month rainy season. For the Amazon forest evapotranspiration is on the order of 1200-1600 mm yr^{-1}. A comparison of precipitation (P), and evapotranspiration (E), for oceans, continents, and the Amazon Basin demonstrates that the high Amazon evapotranspiration rate is much more like those of oceans than of average continental situations.

OCEANS	P $\simeq$ 1,100 mm yr^{-1} E $\simeq$ 1,200
CONTINENTS	P $\simeq$ 710 mm yr^{-1} E $\simeq$ 470
AMAZON	P $\simeq$ 2,300 mm yr^{-1} E $\simeq$ 1,200 - 1,500 - 1,600

The Penman methodology (Penman, 1963; DeBruin, 1983) for estimating evapotranspiration can be applied to different vegetation types including, with some modification, forest. There has been a consequent tendency to believe that combined evaporation and evapotranspiration should recycle water to the atmosphere to the same extent whatever the type of vegetation, whether forest, pasture, or annual crops. This does not take into account the increased surface run-off and reduced water recharge of the soil caused by deforestation. If there is high evapotranspiration following deforestation it is likely to be drawing from a decreased soil water store.

Water consumption from bare soils and partial vegetation cover is between 400 and 500 mm yr^{-1} (Baumgartner, 1979). Molion and Bettancurt (1981) demonstrated that evaporation difference among types of vegetative cover can be explained quantitatively in terms of energy considerations, assuming water is not limiting. They show forests evaporate more water than any other cover type, as much as twice that from bare soil. Tree roots penetrate a large volume of soil and to greater depths than do roots of most annual crops and replacement vegetation, and water recharge of forest soils is substantial. If forest is cleared and precipitation remains the same in the short run, the decrease in evapotranspiration will lead to increased run-off.

Deforestation also has a direct effect on run-off. The forest canopy protects the forest floor from direct impact of rain, while surface litter and organic matter promote water absorption. With forest clearing the vegetation no longer maintains soil structure, percolation, and water recharge normal for the systems.

Reduced infiltration rates on newly cleared tropical forest soils is usual. On cleared primary forest areas near Manaus, Schubart (1977) reported infiltration rates on five year old pasture, one tenth those for forest. In West Africa, Charreau (1972) recorded run-off twenty or more times greater for cultivated and bare soils than from forest. Deforestation does not invariably result in increased run-off. In rainforest of Queensland, Australia, Bonnell et al. (1983) found prevailing rainfall intensities (4239 mm yr^{-1}) frequently exceeded saturated hydraulic conductivity of the profile. There is consequently rapid saturation of the top layer followed by generation of surface flow. After deforestation there was little increase in run-off although soil erosion rates increased by a factor of ten (in the absence of vegetation to break the impact).

In general, tropical deforestation seems initially to increase peak run-off of minor river flows, which later stabilize at lower levels. In Malaysia, conversion of natural forest to rubber or oil palm doubled peak storm flows, halved low flows, and greatly increased erosion. In one cachement, low flows declined to one quarter of previous levels (Daniel and Kulasingham, 1974).

WATER BALANCE IN THE AMAZON

Water balance and recycling estimates can be made for the Amazon basin because measurements of Amazon River flow (Oltmann, 1967) show how much water leaves the basin annually as river water. Water budgets have been estimated using meteorological data and radiosonde data, and the recycling confirmed with stable isotope data. This provides essentially three independent lines of evidence for recycling of water within the basin.

All the water that falls as rain in the Amazon does not leave via the river. Molion (1975) estimated that 52% of the rain falling on the basin left via the river. Similarly, Villa Nova et al. (1976) applied Penman's method, as adapted for forested regions by Shiau and Davar (1973), and calculated 46% of the rain was lost to the river and 54% was evaporated in the basin. Allowing for the scale of the system and the inherent inaccuracies, these estimates are reasonably close.

Not enough moisture arrives from the Atlantic to account for all Amazon rainfall. Marques et al. (1977) obtained data for water vapor fluxes and computed precipitable water from the analysis of 658 daily meteorological upper air observations by radiosonde in the vicinity of Manaus and Belém. The study demonstrated that the Atlantic ocean contributed 52% of the regional precipitation and that the rest of the water vapor was supplied from within the basin. This pointed toward a significant role for local evapotranspiration in generation of rain in the area, with 48% of rain contributed by evapotranspirations.

In summary, the Amazon basin receives a total of 11.87×10^{12} m^3 yr^{-1} of rain water and loses 6.43×10^{12} m^3 yr^{-1} through evapotranspiration and 5.45×10^{12} m^3 yr^{-1} through river discharge. Marques et al. (1980) observed that the amount of water vapor coming from the ocean is of the same magnitude as that generated by evapotranspiration. It should be noted that the measurement of water vapor from the ocean is independent from the estimation of evapotranspiration (by rainfall minus river discharge).

Such values refer to the Amazon basin as a whole, only three quarters of which is covered with high tropical forest. Specific values may be expected to vary considerably, and be higher in high tropical forest areas. Two careful studies of specific watersheds with high forest cover near Manaus, where rainfall averages 2000 mm yr^{-1} show local evapotranspiration can be much higher than average. The results of the two studies (Franken et al., 1982; Leopoldo et al., 1984) showed respectively that 18.7% and 25.6% of rain was intercepted by the forest and evaporated to the atmosphere, 62.0% and 48.5% was transpired by the forest, with run-off consisting of 19.3%

and 25.9%. Thus combined evaporation and transpiration represented 80.7% and 74.1% of total rainfall at these two high forest locations.

Systematic measurements of the isotopic composition of Amazon rainwater have been made. They show less of a decrease in levels of O_{18} between oceanic levels at Belém and at points further west than would be expected from continuous removal, through precipitation, of water vapor of solely oceanic origin. This model assumes oceanic vapor as the only source and includes the selective loss of O_{18} as moisture condenses to form precipitation.

The O_{18} levels do fit with another model that assumes rain derives from water vapor originating not only from oceanic vapor, but also from evapotranspiration (which selectively takes up O_{18} relative to other isotopes of oxygen) (Dall'Olio et al., 1979; Salati et al., 1979). The model assumes a predominant flow from east to west (Newell et al., 1972). Vapor influx from the Atlantic was estimated at Belém from radiosonde measurements of wind and humidity. A central region of the basin between 0° and 5° S latitude was arbitrarily subdivided into eight 3° segments between Belém (48°30') and just west of Benjamin Constant (72°30). The model assumes that in any one segment, rainfall is about half derived from evapotranspiration within the segment, and half from water vapor conveyed by prevailing winds from the neighboring eastern segment.

It was necessary to have another source of water vapor to explain isotopic data. Evapotranspiration is the most probable. In this case rain is formed by a mixture of water vapor from evapotranspiration and water vapor from the ocean.

Unusually low O_{18} values are sometimes found throughout the Basin during April and May, and occasionally during the late summer months of December-March. These are not due to any large inland isotope gradient, but appear to be imposed on the basin from the outside. During this period the Intratropical Convergence Zone (ITCZ) moves south into the latitudes of the Amazon basin, and the consequent vertical mixing and resultant cloudiness affect isotopic composition and rain patterns. Satellite imagery show these seasonal shifts in wind pattern which must be linked to the isotope data (Salati et al., 1979).

Thus, although on average about half the water vapor in the basin is recycled, external factors (apart from the regular water vapor from the North Atlantic) do exert a major influence on the system from time to time. There is also an efflux of water vapor. Marques et al. (1979, 1980) determined from radiosonde studies that vapor is exported south to Central Brazil and to the Chaco of Paraguay almost every month, but principally in March and December.

POTENTIAL EFFECTS OF AMAZON DEFORESTATION

Deforestation of the Amazon is taking place at a much greater rate than the replacement of the forest, either through replanting with "economic" trees or by comparable systems. Large scale deforestation and possible effects on the hydrological cycle have been discussed recently (Salati et al., 1983; Salati and Vose, 1983a, 1984, 1984a; Gentry and López-Parodi, 1980, 1982; Nordin and Mead, 1982). The speed of deforestation is currently officially estimated at a minimum of 2.3×10^6 ha yr^{-1} (Anon. PMCFD, 1982) in the Brazilian Amazon.

There is some evidence that the rate and extent of deforestation is greater than officially recognized. Official data, based on satellite photographs taken in 1975 and 1978 indicated only 1.55% of legal Amazonia had been deforested. These studies represented a first attempt to estimate the rate of deforestation, but did not distinguish between primary and replacement forest. In some areas, like the Bragantina region east of Belém earlier colonization had cleared areas larger than the official estimate. Furthermore, Legal Amazonia is only about half rainforest, so if the percent is taken against forest area, it would be about double that given against Legal Amazonia.

More recent data (Fearnside, 1982), based on satellite photographs up to 1980, showed an exponentially increasing rate of deforestation, especially in the southwest of the Brazilian Amazon (Mato Grosso, Rondônia, and Acre). About 3% of the new (1979) state of Mato Grosso has been deforested from 1978 to 1980. In other regions, such as the states of Pará and Amazonas, the increase, while remarkable, has had a more linear tendency. Local observation by researchers on scientific trips subsequent to 1980 indicate a great pressure on the forest and large areas replaced by agricultural activities and pastures. Actual current deforestation for the entire Amazon could be as much as 4.0×10^6 yr^{-1}.

Without adequate and almost immediate replacement by other vegetation, deforestation may result in environmental modification through soil damage and erosion, alteration in biogeochemical cycles, increased water run-off and loss, with change in the micro- and possibly meso-climate.

Deforestation on a large scale will alter the short term energy balance. A large part of the energy used by forest plants in the process of transpiration will instead heat soil and air. Thus the water balance is inevitably related to the energy balance.

Deforestation will affect the water cycle directly because of greater surface run-off. The less water available for evapotranspiration will decrease relative air humidity and alter energy balance. Incident solar energy will heat air instead of going into water evaporation. The findings by Ribeiro and Santos (1975) of higher temperature for campinas (low woody vegetation on white sand) than in neighboring forest near Manaus are consistent with this.

Salati et al. (1983a) noted there probably would be little measurable effect in small clearings because the air would be swamped by moist air surrounding forest. Nonetheless, water recycling will have been reduced by a measurable, if small, amount and the effect of a large number of clearings would be large in the aggregate. Deforestation of large areas (between 1 and 100 km^2) will probably result in significant reduction in water being returned to the atmosphere, diminished local cloud cover and consequently increased solar radiation. Precipitation arising from local recycling would therefore be reduced probably.

Reduced precipitation could have profound effects for the ecology of natural systems, cultivation, and all biologically based enterprises. The potential effects of Amazon deforestation on a wider scale are harder to assess. Large changes in vegetation cover in these regions might affect regional climates. Attempts to model potential climatic effects of Amazon deforestation include a surface energy balance model (Lettau et al., 1979) and a two dimensional zonal atmosphere model (Potter et al., 1975, 1981). The latter study suggests a decrease in rainfall of 230 mm yr^{-1}. This is, of course, only an initial estimate. There is an important energy transfer during evaporation: solar energy is transformed into latent heat which is then released in the highest layers of the atmosphere during the formation of clouds through vapor condensation. This energy contributes to circulation of the upper atmosphere as well as to transfer of energy from equatorial to polar regions.

Dickinson (1981) concluded complete tropical deforestation is not likely to cause global climatic changes greater than natural ones. At the present time it seems fairly certain that Amazon deforestation would cause major changes within the Basin itself. If the reduction in rainfall was uniform throughout the year and throughout the Basin then little practical effect might be anticipated because the current estimated decline is on the order of 10% or less. It is unlikely the decline will be distributed evenly in time or geographically. Rather it will be greater in those places and at those times where local recycling is presently of greatest significance. In the region of Manaus the present dry period (as much as 73 days) is already close to the maximum the ecosystem can withstand, and any lengthening of the dry period or reduction in precipitation at other periods, might induce irreversible changes.

The situation of forests such as those in the vicinity of Manaus indicates that they are vulnerable to even small reductions in humidity, whether generated by local deforestation or a more general drying trend. Already, marked changes have been noted in the woody plant communities in small forest remnants (10 ha) left in the course of deforestation just north of Manaus (Lovejoy et al., 1984). These changes are certainly correlated with, and are highly likely to be caused by fluctuations in air temperature and humidity generated by the surrounding pasture land even where there is no evidence of reduced rainfall.

The insulating capacity of rain forests in maintaining very constant temperature and humidity conditions is very great. When Allee (1926) took such measurements over half a century ago in Panama he noted better temperature control than in his thermostatically controlled office at the University of Chicago. This capacity is disrupted by the creation of forest margins. While the immediate effect of microclimatic change induced by habitat fragmentation is not strictly comparable to the effects of reduced precipitation (because among other things the growth of secondary vegetation should restore insulation around forest fragments) it does give an indication of the sensitivity of these forests to reduced moisture.

The changes related to the edges of the reserves isolated by pasture in the Minimum Critical Size of Ecosystems project are by no means limited to the woody plants (Lovejoy et al., in press). The latter can be taken as an indicator of gross changes in the animal community, some generated by physical factors, and some as second order effects by the changed woody plant community. To the extent these results can be taken as indicative of what reduced rainfall might engender, and especially in conjunction with habitat fragmentation, a large number of extinctions of endemic species might ensue.

One might ask how the central Amazon ever came to be forested as it is today, when it is likely (Haffer, 1969; Prance, 1973, 1982) that rain forest did not exist there during Pleistocene interpluvials? It certainly appears plausible that the necessary rainfall--and vegetation-- generated hydrological cycle could have been approached incrementally. Alternatively, the forest could have developed during a period of higher precipitation and may persist today under almost marginal conditions only because of its capacity to contribute to the hydrological cycle.

Gentry and López-Parodi (1980, 1982) have suggested there are already indications of increased run-off due to deforestation in the upper Amazon which may lead to increased frequency and degree of flooding in the lower Amazon, but this has been questioned (Nordin and Mead, 1982). Certainly reduced precipitable water in the Basin would reduce water available for export to the Chaco of Paraguay and Central Brazil (and lower rainfall in those areas). A drier regime in the Amazon and lowered rainfall in Central Brazil could increase continentality, influencing climatic patterns, e.g., colder or longer winters in south-central Brazil, and affecting such crops as oranges and sugarcane.

FUTURE RESEARCH ON THE AMAZON SYSTEM

Apart from the national work of INPA (Instituto Nacional de Pesquisas da Amazônia), EMBRAPA (Empresa Brasileira de Pesquisa Agropecuaria) and CENA (Centro de Energia Nuclear na Agricultura, University of São Paulo), there are a number of current external contributions to Amazon natural systems research. For example, the Max Planck Institute is working on limnology and nutrient cycling, a University of Washington team supported by the National Science Foundation (U.S.) is studying the biogeochemistry of the River Amazon, the Institute of Hydrology (U.K.) is working on micrometeorology of the high forest, plus others, all in collaboration with Brazilian counterparts. The World Wildlife Fund (U.S.) is supporting studies to determine the minimum area of forest necessary to preserve plant and animal ecosystems.

Despite all the work that is in progress, there is still an urgent need to carry out baseline research on natural systems, so that in the future it may be possible both to detect the changes that will result from changing land use, and maybe also warn us of some of the possible consequences. At the present time INPA, EMBRAPA, and CENA are actively working together to establish a large and comprehensive international program "to assist through research, data acquisition and training the planned use of Brazilian Amazon renewable natural resources on the basis of improved knowledge of the relevant ecological and hydrological problems, and to establish a data base."

Six areas of research, all interrelated, have been identified for specific study: (1) the hydrological cycle; (2) primary production studies; (3) carbon cycle; (4) nitrogen cycle; (5) soil fertility and nutrient cycle; and (6) agrochemical use, degradation, and transformation. The objectives will be:

● Study quantitatively water, nitrogen, carbon dioxide and other nutrient cycles.

● Seek better identification of the origin of the water vapor producing precipitation and to establish better water vapor circulation models, to ascertain the relative importance of water vapor stemming from evapotranspiration, and from other sources, for the water economy of the region.

● Establish, where needed, physical and biological "base line data" for the undisturbed soil and water ecosystems.

● Establish rates of evapotranspiration and run-off from the vegetation of a variety of land uses. Similarly to indicate the changes to be expected in species abundance, diversity, and succession, and in the levels of important agricultural nutrients as a result of different methods of forest clearing, plant residue conservation, and subsequent agricultural practice.

● Quantify the changes to be expected in the quality of river and other derived water bodies as a result of forest clearance, and agricultural substitution and development, e.g., effects of increased run-off.

● Establish mathematical models for describing the changes, hydrological and nutrient cycles, and for predicting trends or changes expected in future.

● Develop guidelines for the protection of soil structure and essential plant nutrients under the conditions of required crop production.

● Develop guidelines for the protection of the quality of river and lake and other water bodies and their abundance where these may be threatened.

● Make available information and data that are essential for agricultural development of the region, for the protection of its natural resources, and for ensuring long-term agricultural stability.

Clearly this is a very large and ambitious program which will take many years to fulfill, although a start has been made with existing programs.

REFERENCES

Allee, W. C., 1926: Measurements of environmental factors in the tropical rain forest of Panama. Ecology 7: 273-302.

Anon: Alteração da cobertura vegetal natural da Região Amazônica. Programa de monitoramento da cobertura vegetal do Brasil (PMCFB). Mimeo Table, Brasília December (1982).

Baumgartner, A., 1979: Climate variability and forestry. In World Climate Conf. Proc. WMO, Geneva.

Bonnell, M., D. A. Gilmour and D. C. Cassells, 1983: Runnoff generation in tropical rain forests of northeast Queenland, Australia and the implications for landuse management. Pages 287-197. In: R. Keller (ed.), Hydrology of Humid Tropical Regions. IAHS Pub. 140.

Charreau, C., 1972: Problemes posés par l'utilisation agricole des sols tropicaux par des cultures annuelles. Agron. Tropical (France) 27: 905-929.

Cochrane, T. T. and P. G. Jones, 1981: Savannas, forests and wet season potential evapotranspiration in tropical South America. Trop. Agr. (Trinidad) 58(3): 185-190.

Dall'Olio, A., E. Salati, C. T. Azevedo, and E. Matsui, 1979: Modelo de fracionamento isotópico da agua na bacia Amazônica. Acta Amazonica 9(4): 675-687.

Daniel, J. G. and A. Kulasingham, 1974: Malayan Forester 37: 152.

DeBruin, H. A. R., 1983: Evapotranspiration in humid tropical regions. Pages 299-311. In: R. Keller (ed.), Hydrology of Humid Tropical Regions IAHS Pub. 140.

Dickinson, R. E., 1981. Studies in Third World Societies

14:    411, Publ. William and Mary College (Williamsburg, VA).

Fearnside, P. M., 1982: Deforestation in the Brazilian Amazon: How fast is it occurring? Interciencia 7(2): 82-88.

Franken, W., P. R. Leopoldo, E. Matsui and M. N. Goes Ribeiro. 1982. Estudo da interceptação da agua de chuva em cobertura florestal Amazônica do tipo terra firme. Acta Amazonica 12(2): 327.

Franzle, O, 1979: Appl. Sci. and Develop. 13: 88 (Inst. Sci. Coop., Tubingen, Germany).

Gentry, A. H. and J. López Parodi, 1980: Deforestation and increased flooding of the upper Amazon. Science 210: 1354-1356.

_____, 1982: Deforestation and increased flooding of the Upper Amazon. Science 215: 427.

Haffer, J., 1969: Speciation in Amazonian forest birds. Science 165: 131-137.

Leopoldo, P. R., W. Franken and E. Matsui, 1984: Hydrological aspects of tropical rain forest in central Amazon. 44th Int. Congr. Americanists: Change in the Amazon. Manchester Univ., U.K. 5-10 Sept., 1982. Manchester Press.

Lettau, H., K. Lettau and L. C. B. Molion, 1979: Amazonia's hydrologic cycle and the role of atmospheric recycling in assessing deforestation effects. Mon. Wea. Rev. 197: 227-238.

Linsley, R.K., 1951: The hydrologic cycle and its relation to meterology. Pages 1044-1054. In: T. F. Malone (ed.), Compendium of Meteorology, Amer. Meter. Soc.

Lovejoy, T. E., J. M. Rankin, R. O. Bierregaard, Jr., K. S. Brown, Jr., L. H. Emmons, and M. E. Van Der Voort, 1984: Ecosystem decay of Amazon forest remnants, pages 295-325. In: M. H. Nitecki (ed.), Extinctions. Univ. Chicago Press, Chicago.

Lovejoy, T. E., R. O. Bierregaard, Jr., A. B. Rylands, J. R. Malcolm, C. E. Quintela, L. H. Harper, K. S. Brown, Jr., A. H. Powell, G. V. N. Powell, H. O. R. Schubart, and M. B. Hays. in press. Edge and other effects of isolation on Amazon forest fragments. In: Conservation Biology II, Sinauer Associates, Sunderland, MA.

Marques, J., J. M. dos Santos, N. A. Villa Nova, and E. Salati, 1977: Precipitable water and water vapour flux between Belém and Manaus. Acta Amazonica 7(3): 355-362.

Marques, J., J. M. dos Santos, and E. Salati, 1979: O

campo do fluxo de vapor dágua atmosférico sobre a região Amazônica. Acta Amazonica 9(4): 701-713.

Marques, J., E. Salati, and J. M. dos Santos, 1980: A divêrgencia do campo do fluxo do vapor d'água e as chuvas na região Amazônica. Acta Amazonica 10(11): 133-140.

Molion, L. C. B., 1975: A climatonomic study of the energy and moisture fluxes of the Amazonas basin with consideration of deforestation effects. Ph.D. Thesis, University of Wisconsin, U. S. A.

Molion, L. C. B. and J. J. U. Bettancurt, 1981: Land use in agrosystem management in humid tropics. In: J. J. Talbot and W. Swanson (eds.) Woodpower, New Perspectives on Forest Usage. Pergamon Press, Oxford and New York.

Newell, R. C., J. W. Kidson, D. G. Vincent and G. J. Boer, 1972: The General Circulation of the Tropical Atmosphere. MIT Press, Cambridge, Mass Vol. I.

Nordin, C. F. and R. H. Mead, 1982: Deforestation and increased flooding of the Upper Amazon. Science 215: 427.

Oltmann, R. E., 1967: Reconnaissance investigations of the discharge and water quality of the Amazon. Pages 163-185. In: H. Lent (ed.) Atas do Simpósio sobre a Biota Amazônica 3 (Limnologia).

Penman, H. L., 1963: Tech. Comm., No. 53, Commonwealth Agr. Bureaux, Farnham Royal.

Pereira, H. C., 1973: Land Use and Water Resources in Temperate and Tropical Climates. The University Press, Cambridge, MA, 146 pp.

Potter, G.L., H. W. Ellsaesser, M. C. McCracken and F. M. Luther, 1975: Possible climatic impact of tropical deforestation. Nature 258: 697.

Potter, G. L., H. W. Ellsaesser, M. C. McCracken and J. S. Ellis, 1981: Albedo change by man: Test of climatic effects. Nature 291: 47.

Prance, G. T., 1973: Phytogeographic support for the theory of Pleistocene forest refuges in the Amazon basin, based on evidence from distribution patterns in Caryocaraceae, Chrysobalanaceae, Dichapetalaceae and Lecythidaceae. Acta Amazonica 3(3): 5-28.

Prance, G. T. (ed.) 1982: Biological Diversification in the Tropics. Columbia Univ. Press, New York, 714 pp.

Ribeiro, M. N. G. and J. M. dos Santos, 1975: Observações microclimáticas no ecossistema Campina Amazônica. Acta Amazonica 5: 183-187.

Salati, E., A. Dall'Olio, E. Matsui and J. R. Gat, Jr.,

1979:   Recycling water in the Amazon basin: An
isotopic study.   Water Resources Research 15(5):
1250-1258.

Salati, E, T. E. Lovejoy and P. B. Vose, 1983:
Precipitation and water recycling in tropical rain
forests with special reference to the Amazon Basin.
Environmentalist 3(1): 67-72.

Salati, E. and P. B. Vose, 1983a: Depletion of tropical
rain forests. Ambio 13(2): 67.

Salati, E. and P. B. Vose, 1984: in press. The water
cycle in tropical forests, with special reference to
the Amazon. Papal Acad. of Science, Symp. on Chemistry
of Atmosphere, Rome, 7-11 Nov., 1983.

Salati, E. and P. B. Vose, 1984a: Amazon basin: a system
in equilibrium. Science 225: 129-138.

Shiau, S. V. and K. S. Davar, 1973: Modified Penman method
for potential evapotranspiration from forest regions.
J. Hydrol. 18: 349-365.

Schubart, H. O. R., 1977: Criterios ecologicos para o
desenvolvimento agricola das terras firmes da Amazonia.
Acta Amazonica 8(4): 559.

UNESCO, 1971: Scientific Framework of World Water Balance.
UNESCO Sc. 70/XXI: 7/A, Geneva

Villa Nova, N. A., E. Salati and E. Matsui, 1976:
Estimativa da evapotranspiração na bacia Amazonica.
Acta Amazonica 6(2): 215.

WMO, 1983: Operational hydrology in the humid tropical
regions. Pages 3-26 In: R. Keller (ed.), Hydrology of
Humid Tropical Regions. IAHS, Publ. 140.

6. Catastrophic Drought and Fire in Borneo Tropical Rain Forest Associated with the 1982-1983 El Niño Southern Oscillation Event

ABSTRACT

Drought and fire devastated over five million ha of lowland rain forest in north and east Borneo in 1983. Measures collected before and after the disturbance from one burned primary forest site established that approximately 25% of the canopy trees were killed, mostly from drought, while over 90% of both understory treelets and lianas were killed, mostly from fire. Mortality was 40-60% among canopy trees producing fruits and seeds important to vertebrates, and crown dieback affected the majority of surviving fruit trees. Hornbill and squirrel species populations have declined, but those of several primates have remained approximately stable. Analysis of long-term rainfall data indicates that a drought of this severity occurs perhaps on the scale of once each generation of canopy trees, but that the accompanying large-scale fires may have resulted from recent anthropogenic changes and may be without historical precedent.

INTRODUCTION

The climatic perturbations of the 1982-83 El Niño - Southern Oscillation (ENSO) event were the most severe in at least a century (Cane, 1983; Barber and Chavez, 1983; Kerr, 1983), and engendered large-scale disturbances that grossly altered the species compositions and structures of ecological communities. Massive dieoffs of dominant marine organisms accompanied extraordinarily widespread, severe warming of Pacific waters (Dayton and Tegner, 1984; Schreiber and Schreiber, 1984). Although it has been reported that severe drought affected subtropical woodland and agricultural systems (Canby, 1984), effects on tropical rain forests have until now gone unreported.

A severe drought gripped some tropical forest regions of Southeast Asia from July 1982 through April-May 1983. Most affected were the Malaysian state of Sabah in northern Borneo and particularly the Indonesian province of East Kalimantan in eastern Borneo. Fires originating with slash and burn agriculturalists combined with drought to devastate a vast region of lowland evergreen rain forest. The acreage affected in Sabah is not accurately known but is apparently large (1-2 million ha) judging by information obtained in Sabah (Beaman, et al., mss.) and from the examination of NOAA-7 satellite images (J. P. Malingreau et al., 1985). Approximately 3.7 million ha were severely damaged in Kalimantan (Lennertz and Panzer, 1983; Wirawan, 1984). This area comprised 800,000 ha of primary lowland rainforest, 550,000 ha of peat swamp forest, 1,200,000 ha of selectively-logged primary forest, and 750,000 ha of shifting cultivation and settlements (Fig. 6.1) (Lennertz and Panzer, 1983). Damage from fire was much more severe in selectively-logged primary forest, where woody residues left from logging fueled hot fires (Lennertz and Panzer, 1983; Kartawinata et al., 1980; Leighton, 1984; Wirawan, 1984). The cooler, slower burning fires in primary lowland forest were fueled by leafy litter (Lennertz and Panzer, 1983; Leighton, 1984; Wirawan, 1984; and L. Berenstain, pers. comm.). Highest tree mortality and structural damage is thought to have occurred in peat swamp forest, because a meter or two of the dried peat soil burned, destabilizing trees which were then easily toppled by wind. Both satellite remote sensing data (Malingreau, et al., 1985) and aerial surveys by small plane have established that most of this area burned. Some large areas (ca. 100,00 ha) within the affected region did not burn, although trees were killed by drought (Leighton, 1984: Wirawan, 1983, 1984; J. P. Malingreau, pers. comm.).

Fires within East Kalimantan were restricted to the relatively drier lowland areas of rain forest (Fig. 6.1). Rainfall was only a third of normal during the 10 months of drought (e.g., Mentoko 34.9%, Samarinda 23.4%, Kota Bangun 31.5%, and Balikpapan 39.5% of normal) (Fig. 6.1). The drought of 1982-83 was unusually severe because the normal August-October dry season was extended and extreme, and followed by an extraordinarily hot, cloudless, and rainless February, March, and April (Fig. 6.1). Rainfall at four widely spaced stations (Fig. 6.1) averaged only

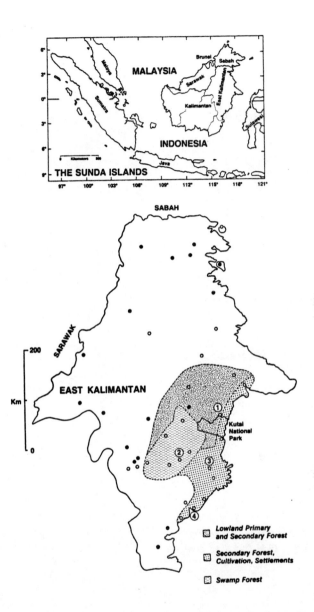

Fig. 6.1(a) (Top) Map of Sundaland, showing location of East Kalimantan. (Bottom) Extent of burned area and rainfall in East Kalimantan, indicating different vegetation types affected. Solid circles are stations where mean annual rainfall exceeds 2500 mm. Circled numbers indicate location of measurement stations mentioned in Fig. 6.1(b).

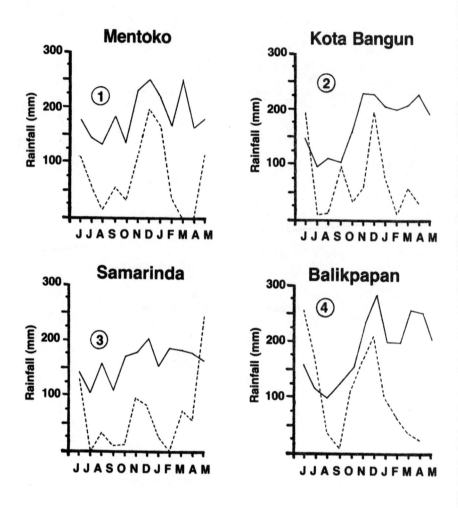

Fig. 6.1(b) Long-term mean monthly rainfall (solid line) versus rainfall from June 1982 through May 1983 (dashed line) for four localities on map (see Fig. 6.1 a), cross-referenced by number. (Adapted from map in Lennertz and Panzer, 1983; see also Malingreau et al., 1985; compiled from aerial survey.)

15.1% of normal for these three months. As a rule of thumb, tropical evergreen rain forest suffers drought stress when monthly rainfall falls below approximately 100 mm (Whitmore, 1984), and these stations averaged 13-43 mm per month. In response to drought stress, many of the normally evergreen trees and lianas of this forest dropped most of their leaves during this period, forming a deep, dry combustible litter layer. Fires originating from slash and burn agriculturalists escaped into adjacent primary and secondary forests and burned unmonitored and out of control sporadically during the drought, but especially in March and April, 1983 (Asia Week, 1984; Lennertz and Panzer, 1983; Leighton, 1984; Wirawan, 1983, 1984; Transmigration Area Development, 1983; Malingreau, et al., 1985).

This environmental catastrophe is significant not only because it has disrupted local human communities and engendered great economic losses, e.g. an estimated $US 6 billion loss in future harvestable timber alone (Lennertz and Panzer, 1983), but also because it affected an evergreen rain forest that was thought to be immune to disturbance of this kind. Here we report the effects of this drought and fire on forest structure and describe the mortality among plants and vertebrates that occurred in one primary rain forest site in east Borneo. We also link ENSO events to rainfall patterns in east Borneo and use this and other evidence to address the issue of the historical incidence of severe episodes of drought and fire in this region.

EFFECTS ON VEGETATION IN PRIMARY FOREST

Data on plant and vertebrate mortality in burned primary forest was collected by one of us (ML) at the Mentoko Research Station (0°24',117°6'E) (Rodman, 1978) along the north border of the Kutai National Park (Fig. 6.1). Most of the Mentoko site is comprised of well-drained lowland (30-200 m elev) mixed-dipterocarp evergreen rain forest on sandstone and mudstone (Fig. 6.2), and is representative of the richest rain forests of the world (Whitmore, 1984). The drier ridge and slope forest, dominated by dipterocarps of majestic stature, interdigitates with strips of riparian forest on wetter alluvial soils, which is dominated by Bornean Ironwood (Eusideroxylon zwageri T. & B.). The advantage of studying this site was that the composition, densities, and fruiting phenologies of woody plants, and the densities and diets of

Table 6.1

Mortality among adult fruit trees, and foliage condition of survivors, grouped by primary vertebrate consumer of fruit flesh. n's are individuals fruiting during 1977-1979 that were recensused along trails in September 1983. Asterisks indicate significantly higher mortality when compared with the entire set of equivalently-sized canopy tree using Chi-square tests (see text). Data on condition of survivors was lumped for classes A and B (presented with class B), and not calculated for classes C and F because of small n's.

Fruit class	% Mortality(n)	Condition of Surviving Trees (% of total)			
		Full crowns	Crown dieback	Limbs/Trunk only	n
A. All hornbill-fruits	52 (87) *				
Meliaceae only	62 (52) *				
Myristicaceae only	47 (17)				
B. Other lipid-rich bird-fruits	54 (172)*	24	48	28	103
Lauraceae only	57 (75) *				
C. Sugar-rich bird fruits	55 (64) *				
D. All primate-fruit trees	44 (107)*	45	41	14	56
Anacardiaceae only	25 (24)				
All primate-fruit lianas	97 (61)				
E. All bat-fruits	26 (77)	36	50	14	55
F. Figs (Ficus spp.)					
subg. Urostigma (Stranglers)	83 (86)				
sect. Rhizocladus (climbers)	100 (22)				

diurnal fruit flesh- or seed-eating birds and mammals, (especially primates, squirrels, and hornbills) had been studied at Mentoko for two years during 1977-1979 (Fig. 6.2) (Leighton and Leighton, 1984). Plant mortality was measured by NW in unburned but drought-affected primary forest on similar topography and soils approximately 20 km WSW of Mentoko to separate the effects of drought alone from those of drought plus burn. Data were collected in early September 1983 at the two sites. ML resampled the Mentoko site in August 1984.

Fire entered the Mentoko forest on April 21, 1983 and burned through the litter layer, taking 10 days to traverse the research site (L. Berenstain, pers. comm.). Attesting to the relative coolness of the flames, four months later virtually all stems, including those of even the smallest saplings, remained standing, but dead (Fig. 6.2), a pattern common at other burned primary forest sites (Lennertz and Panzer, 1983; Leighton, 1984; Wirawan, 1984). Occasionally fire climbed up a tree fueled by its load of lianas, and some trees with flammable resins burned. Generally only a surface singe to the bark at the bases of some trees and blackened ground signaled that a fire had occurred. Charred wood was rare. Approximately 5% of the Mentoko forest, comprised of 3-20 m wide discontinuous ribbons of forest on some of the wetter soils bordering perennial streams, escaped burning.

Trees and lianas were measured for stem diameter at breast height (dbh) and classed as dead or alive within vegetation plots at both the burned and unburned primary forest sites (Table 6.1). Classification of trees and lianas as dead or alive was based on whether the plant had live foliage or not. We tested this assumption first by cutting into the bark and wood of dozens of plants without foliage and verifying that all were dead. Second, we observed that foliage sprouting from the trunks or limbs of surviving plants had substantially grown out by early September, indicating that the surviving trees that were leafless at the time of the fire sprouted out with little delay after rains began in June, and that those not immediately leafing out were dead. These assumptions were substantiated during the second year's research, since no individuals listed as dead the first year were found to have recovered. We consulted in advance to ensure that the criteria for classifying dead versus live stems and foliage conditions were identically applied. The "dead" plants censused may have died from causes unrelated to the drought or fire, but this undoubtedly accounts for a miniscule

Table 6.2
Percentage mortality among woody plants in burned and unburned primary forest, Kutai National Park, measured September, 1983

Sample[a]	Site Characteristics[b]	Lianas 4 cm	Percentage mortality (number of individual plants censused) DBH Class of Trees					
			4 cm	4-10 cm	11-18 cm	19-30 cm	31-50 cm	50 cm
A. Burned primary forest								
Plot #1	wet, 70% burned	90.9 (22)	67.8 (242)	47.2 (81)	11.1 (18)	14.3 (42)	25.0 (20)	6.2 (16)
Plot #9	dry, 100% burned	95.0 (20)	83.2 (286)	81.8 (110)	45.4 (33)	37.3 (51)	10.0 (20)	23.3 (30)
Plot #20	dry, 100% burned	95.1 (41)	89.8 (147)	64.6 (96)	45.0 (20)	42.9 (28)	31.6 (19)	21.7 (23)
Plot #23	dry, 100% burned		91.5 (247)	65.7 (178)	52.2 (23)	35.2 (54)	41.7 (24)	16.7 (24)
Transect AW26	wet, 80% burned	73.0 (156)			28.4 (116)	18.5 (92)	15.6 (45)	16.7 (36)
Transect CS7	dry, 100% burned	90.0 (118)			51.7 (207)	39.2 (102)	29.5 (61)	17.1 (35)

a Burned primary forest samples: Four plots were randomly selected (after stratification into near and far from river) from among the ten, 0.5 ha plots within the Mentoko site that had been sampled for phenology and forest structure during 1977-1979. All trees 19 cm dbh (diameter at breast height) were censused for mortality within the entire plot (50 m x 100 m). In the sw quarter of each plot all stems 3 cm dbh were censused, and in a randomly selected 10 m x 20 m burned area within this subplot, all woody stems (some only 2 mm diameter) were censused. Transect samples were 10 m x 500 m belts following an E-W compass bearing, running perpendicular to ridges and streams, and thereby sampling habitats in proportion to their occurrence. AW26 was mostly in flat alluvial forest, within 30-100 m of and parallel to the Sangata river. CS7 was parallel to AW26, but 250 m upslope to the south. Unburned primary forest samples: Methods of measurement and categorization of dead and alive trees identical to above samples. Plots were 50 m x 40 m, located approximately 20 km SW of Mentoko within the Kutai National Park, on similar soils and topography. Lianas were not sampled, and n's are unavailable (but would be similar to those of burned plot samples).

Table 6.2 (continued)
Percentage mortality among woody plants in burned and unburned primary forest, Kutai National Park, measured September, 1983

Sample[a]	Site Characteristics[b]	Percentage mortality (number of individual plants censused)				
		Lianas	DBH Class of Trees			
		5 cm	4-10 cm	11-30 cm	31-60 cm	60 cm
B. Unburned primary forest						
Plot #1	dry, 100% unburned	23	20	24	43	71
Plot #2	dry, 100% unburned	n/a	n/a	25	48	37
Plot #3	wet, 100% unburned	6	16	14	0	11

b Wet sites are riverine, on wetter soils; dry sites on ridge slopes or tops. % burned refers to area within plot.

fraction of the mortality (1%?). The dead stems censused had complete branch and twig systems intact, indicating that they had recently died. Also, dead stems, even of seedlings, were rare in this forest prior to the drought and fire (pers. obs.).

Calculating percentage mortality by counting dead stems worked because the fire virtually never burned up even the tiniest sapling. Note, for instance, the high densities of stems 4 cm dbh in each of the 10 m x 20 m subsamples within each plot (Table 6.2). These densities (1.15 individuals per m^2) were similar to the densities the forest supported prior to burning (Leighton, unpubl.). The great majority of saplings were erect but had leafless dead stems, not yet decomposed by insect or fungus attack. Fortunately then, this study was conducted after live stems had flushed leaves again, but before dead ones had fallen.

Comparisons between the two sites in short-term mortality among trees of different sizes four months after the fire established the following (refer to Table 6.2):

i) Over 90% of the large lianas (4 cm stem diameter, those most productive and important as food sources for animals) were killed on burned sites. (The AW26 transect contained more unburned forest patches, so liana mortality was lower).

ii) Over 90% of small trees (5 cm dbh) in burned primary forest were killed. Most of this mortality was attributable to fire, since small trees in unburned primary forest plots experienced on average only a quarter of this mortality.

iii) Large trees had lower mortality in burned forest, and mortality was due primarily to drought. Mortality among large trees (30 cm dbh) in burned forest can be ascribed entirely to drought, since equivalent or higher mortalities were measured in unburned primary forest on similar sites, and few large trees had burned.

iv) Very large trees on dry ridges and slopes were particularly susceptible to drought stress. Seventy-one % and 37% of trees of the largest size class were killed in the unburned ridge samples. In five 500m x 10m belt transects along diferent trail sections at Mentoko, mostly within dry ridge forest, trees 50 cm dbh averaged 45.2% mortality (range of 39.2-60.0%, N=356). These results are consistent with other reports of the susceptibility to drought of large trees on dry sites (Seth et al., 1960; Tang and Chong, 1979).

v) Tree mortality on wetter alluvial soils was roughly half that of drier sites. This was probably due to a

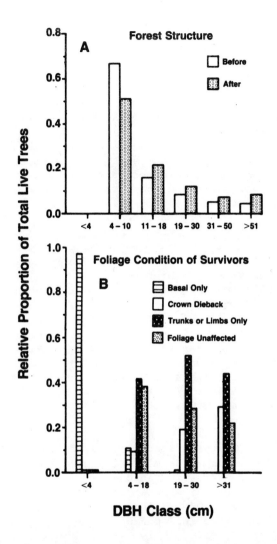

Fig. 6.3(a) Changes in mean relative proportions (n=4 plots) of live trees $\geq$ 4 cm dbh, following drought and fire. Relative proportions of each size class are means of changes in proportions in matched before and after samples from four plots at Mentoko. (b) Relative proportions of surviving trees of different size classes with either normal full crowns of foliage unaffected by fire, truncated crowns due to dieback of terminal twigs, foliage sprouting only from limbs and/or trunk, or leaves sprouting only from shoots at the base of the tree. Data are summed over all individuals from the four vegetation plots.

combination of greater representation by the drought-resistant tree <u>Eusideroxylon</u> <u>zwageri</u> (only 11% of 35 censused trees died, p $<$.01), higher soil moistures and therefore reduced drought stress, and the persistence of small patches of unburned forest in these sites.

vi) Mortality was highest among smaller trees and lianas and lower among larger trees in burned primary forest, whereas the opposite is true on dry sites in unburned forest. As a consequence of the inverse relationship between plant size and mortality in burned forest, the distribution of plant sizes has shifted towards a greater representation of larger trees (Fig. 6.3a). If we were to include stems $<$4 cm dbh in this comparison the pattern would be greatly exacerbated.

The mortality in burned primary forest observed four months after burning can be summarized by combining the samples that are most representative of the proportions of different habitats at Mentoko (i.e. the four plots and transect CS7). By this method, mortality was 91.5% (N=201) for lianas $>$4 cm dbh; 65.7% (N=467) for trees 4-10 cm dbh; 48.2% (N=301) for trees 11-18 cm dbh; 34.7% (N=277) for trees 19-30 cm dbh; 28.5% (N=144) for trees 31-50 cm dbh; and 17.9% (N=128) for trees $>$50 cm dbh. Most small trees and lianas died apparently because their thin bark failed to protect live cambial tissue from the heat of the fire. In contrast, larger trees have thicker outer bark, but their height and high metabolic demands apparently made them more prone to drought stress (Seth et al., 1960; Tang and Chong, 1979).

These mortality figures are from the census of September 1983, and therefore indicate short-term effects of drought and fire. Censuses a year later at Mentoko, however, demonstrated substantial delayed mortality among surviving trees. Of 77 individually-marked canopy trees ($>$30 cm dbh) alive in September 1983, 32% had died within the subsequent year. Only six of these 25 deaths could possibly be attributable to tree falls. The wilted, dead foliage still remaining on many of the dead trees indicated that photosynthetic gains from their meager foliage regrowth following the drought outstripped respirational costs.

Structural damage and tree mortality will continue to occur in the coming years as dead canopy trees fall, further isolating remaining patches of forest canopy. Only 36% (N=61) of the dead canopy trees ($>$30 cm dbh) had fallen within 15 months after the fire, although virtually every 100 m of the trail system was already obstructed by a major

new treefall, often comprised of several canopy trees.
Canopy foliage volume has also been severely reduced
because of extensive drought-induced crown dieback among
trees surviving the fire and drought. Only 30% of the
survivors had a crown foliage equivalent to pre-fire
densities and volumes in September 1983 (Fig. 6.3b). In
40% of the canopy trees, crowns were severly reduced
because terminal twigs had died back, so that foliage
regrowth was restricted to some of the larger twigs or
branchlets (see Fig. 6.2). A further 20% of these trees
were even more severely affected; foliage had sprouted only
from some of their major limbs and/or trunks. In contrast,
the few saplings (<4 cm dbh) surviving in burned forest had
sprouted from the dead stem base (98% of the survivors).

A year later, recensus of 42 of the canopy survivors at
Mentoko indicated that these trees had not recovered
foliage volume; average reduction in crown volume was over
50% (estimated by eye to the nearest 20% of the pre-drought
and fire crown volume of the tree, as assessed from branch
structure). The small, bushy crowns of these survivors
(Fig. 6.2d) will apparently be permanent features, as found
in a study of Indian forest linking these features to
drought stress (Seth, et al., 1960; Tang & Chong, 1979).

In sum, the virtual total loss of lianas from the
burned primary forest at Mentoko, the approximately 25%
mortality among canopy trees, and the substantial loss of
foliage volume by 70% of the surviving canopy trees, has
resulted in a canopy that is grey and open in aspect,
intercepts little radiation, and is incapable of buffering
understory microclimate as does a normal rain forest
canopy. Other changes that are likely to be important to
the animal community include a decline in the availability
of food resources drawn from plant reproductive parts,
destruction of arboreal travel pathways, loss of cover and
shelter from predators and climate (sun, wind, and rain),
and loss in richness of habitat microsites which underlie
the diversity of small organisms in tropical rain forests.

EFFECTS ON FRUIT TREES IMPORTANT TO VERTEBRATES

To document changes in the availability of fruit
resources for various birds and mammals, individually-
marked and identified trees that had fruited during 1977-
1979 along ca. 12 km of the trail system were recensused.
All previously-tagged trees were recensused along four
sections of the trail system, selected without regard to

plant composition or habitat. The seeds or fruits of these trees had been identified previously as dietary items of specific vertebrates (Leighton and Leighton, 1983). Mortality or foliage loss to fruit trees most important to specific types of vertebrates can therefore be distinguished.

The results of these censuses (Table 6.1) show that the fruit trees of primates and birds (especially of hornbills) were disproportionately killed by drought and/or fire in relation to their relative frequencies among all trees of matched sizes and habitats. Also, surviving vertebrate fruit trees suffered foliage losses similar in degree to the overall set of canopy trees (Table 6.1). Fifty-two % of the canopy trees were killed which produce the large, dehiscent, fleshy fruits preferred by and dependent upon seed dispersal by six hornbill species (Leighton and Leighton, 1982; Leighton, 1983). This set of trees provides a continuous supply of nutrient-rich fruits that sustain four of the territorial species during fruit-poor seasons. Trees of Meliaceae, the most important family of these fruits, suffered high mortality (Table 6.1) even though they mostly stand on wetter alluvial sites of relatively low fire damage; it is notable that these trees have thin bark of little insulating power against fire. Also killed were over half the major fruit trees (50+ species of Lauraceae are most important) preferred by other large-bodied frugivorous birds, which include three species of fruit pigeons (Ptilinopinae), six species of barbets (Capitonidae), the hill myna Gracula religiosa, and the green broadbill Calyptomena viridis (Table 6.1).

Altogether 44% of the canopy fruit trees of species fed upon by the frugivorous primates---orangutans (Pongo pygmaeus), gibbons (Hylobates muelleri), and macaques (Macaca fascicularis and M. nemestrina)---were killed (Table 6.1). In contrast to the Meliaceae preferred by hornbills, the two most important primate fruit trees (both Anacardiaceae), Koordersiodendron pinnatum (Blanco) Merr., and Dracontomelon dao (Blanco) Merr. & Rolfe, survived well (Table 6.1). Few Dracontomelon died because they grow on wetter riverine sites; virtually all individuals were undamaged and in fact fruited during August 1984. On the other hand, liana fruit contributes significantly to the total food budgets of primates in this forest, but the most important set, large woody climbers of the Annonaceae, suffered 97% mortality. These annonaceous lianas ripen fruit outside community-wide fruiting peaks, thereby stabilizing fruit availability for primates (Leighton and

Leighton, 1983).

Perhaps most significant of all for vertebrates was the high mortality among figs (Ficus spp.) at Mentoko. Ninety % of the adult figs of the 26 species of strangling or tree-like form (subg. Urostigma) and the 7 species of liana form (subg. Ficus, sect. Rhizocladus) that had fruited in 1977-79 were killed (Table 6.1). These two groups of figs are especially important food sources for dozens of bird and mammal species because fig plants fruit abundantly and aseasonally (Leighton, 1982), thereby providing a supply of large fruit patches outside of the brief fruiting seasons of more preferred fruit types (Leighton and Leighton, 1983).

MORTALITY AND RESILIENCY AMONG FRUGIVOROUS VERTEBRATES

Changes in vertebrate densities were assessed by comparing the rates of encountering vertebrates measured during 1977-79 with those during the 1983 and 1984 periods of fieldwork. Only striking changes are mentioned. (We are grateful to J. Mitani for assessing effects on orangutan and gibbon populations, and to L. Berenstain for assessing effects on long-tailed macaques.)

Given high fruit tree mortality and the loss of productive capacity among surviving fruit trees, it is not surprising that most populations of frugivores have declined in the burned primary forest. Losses of frugivorous birds have been most extreme. For example, two of the five territorial hornbill species are no longer present. The other three species (Anorrhinus galeritus, Anthracoceros malayanus, and Buceros rhinoceros) were sighted or heard rarely and have certainly declined in density. Since alternative appropriate habitat is undoubtedly already filled with other territorial groups, it is rather inconceivable that the missing hornbill groups could have successfully moved and survived elsewhere. The three species of diurnal seed-eating squirrels which were common in the forest prior to burning, Ratufa affinis, Callosciurus prevostii, and C. notatus, have definitely declined in numbers.

Population densities of the large-bodied anthropoid primates have been the least affected of the major arboreal vertebrates. Resightings of individually-known orangutans (eight individuals), gibbons (four families), and long-tailed macaques (one large group) established that these three populations had not declined at Mentoko by

12-16 months after the fire. At least some infants in each of the three populations both survived the fire and were born subsequent to the fire, and were apparently healthy 12-16 months later. The other large arboreal primate, the grey leaf monkey Presbytis hosei, was also commonly resighted but assessment of relative changes in density for it and for the ground-traveling Macaca nemestrina require further study.

The resilience of these primates is remarkable, considering the extensive structural damage to the forest and the loss of fruit and seed resources that these species prefer to eat over other food types. Their adaptability is presumably due to their generalized omnivorous diets and behavioral flexibility which allows them to readily switch food types. These primates switch diets seasonally to subsist largely on leaf, except in the case of orangutans, which subsist on bark and diets of low digestibility when preferred fruits and seeds are relatively rare (Leighton and Leighton, 1983). Details and activity shifts by Macaca fascicularis at Mentoko up to five months after the fire are reported in detail by Berenstain (1985).

Frugivorous birds and mammals which are smaller and less flexible in diet than primates do not switch to leaf or bark eating, and require consistent diets of a high proportion of insects, fruit flesh, or seeds. Judging from how barren of vertebrates this forest appears (woodpeckers feasting on abundant wood-boring beetle larvae excepting), other populations have also apparently declined markedly, but systematic and prolonged censuses are required to confirm these impressions.

PROGNOSIS FOR FOREST RECOVERY

Predictions about future community succession towards reestablishment of tall, closed-canopy forest at Mentoko are necessarily tentative since we have no prior studies upon which to draw of the effects of burning and drought on primary rain forest. After normal rains commenced in June following the drought, seeds of a limited number of fast-growing "pioneer" species germinated (Kartawinata, 1980; Whitmore, 1984; Berenstain, 1985). Within three months, seedlings were distributed throughout the burned forest at a median density of 9 individuals per m^2, standing 20-80 cm tall (M. Leighton, in prep.). A year later this regeneration formed a continuous, dense cover of vegetation on average three meters tall, consisting of a mix of

fast-growing, rapidly-maturing tree and liana species.
This regeneration stabilized the soils somewhat, but the
shallow root systems of these pioneer plant species have
not prevented widespread erosion. Steep slopes and
formally stable stream channels showed continued slippage
even with complete vegetative coverage 16 months after the
fire.

It is possible that many of the less common canopy
hardwood trees have or will become locally extinct. In
drought-stricken but unburned forest, a population in which
adults suffer heavy mortality could in time recover since
most of its seedlings or immatures persist, whereas in
burned forest, most small individuals will die. Many
mature -phase species have seedlings or juveniles
ill-adapted to grow in building-phase vegetation, since
they require shaded and moist conditions for growth
(Whitmore, 1984). Recruitment of such species depends upon
the survival of reproductive adults until the softwood
pioneer species mature. Decades will elapse before the
replacement generation of these slower-growing hardwoods
begin fruiting (Whitmore, 1984). The rarer of these
species may fall to population levels too small to recover.

It is difficult to predict how future changes in the
burned forest will affect vertebrate population numbers.
The fast-growing plants which now dominate the regeneration
produce fruit eaten by small birds and only rarely by
primates, hornbills, or other larger birds and mammals. Due
to the demise of maturing juvenile trees and to the
substantial mortality and crown damage suffered by adults,
there is no hope for the short-term recovery of substantial
fruit production by the hardwood canopy species which form
the bulk of the diets of these large mammalian and avian
species. Frugivorous vertebrates which, unlike primates,
are unable to subsist on plant foods other than fruits, are
likely to remain rare. The prognosis for the return to
pre-fire fig densities is also poor, because the
germination sites on tree limbs required by the most
important group of figs, the hemiepiphytic Urostigma, have
become rarer and less suitable, since they are now less
buffered from desiccation. Vertebrate populations
currently with adequate adult densities, like those of the
primates, may in the future decline if resources are too
poor to support the growth of juveniles or if other
unfavorable ecological changes occur.

Our aerial (Leighton, 1984; Wirawan, 1984) and ground
(Wirawan, 1983) surveys indicate, however, that 40-50%
(80,000-100,000 ha) of the Kutai National Park is comprised

of primary forest that did not burn. Although canopy tree mortality from drought stress was substantial in unburned primary forest (Table 6.1; Leighton, 1984; Lennertz and Panzer, 1983; Wirawan, 1983, 1984), and though this forest will sustain additional structural damage as dead trees fall, it is likely that most primary forest plant and animal species have and will persist. These areas can then serve as refuges for vertebrates and for rare trees and lianas that can in the future recolonize burned forest as it undergoes succession towards physical and biotic conditions favorable for various species.

HISTORICAL INCIDENCE OF SEVERE DROUGHT AND FIRE

In evaluating this disturbance in terms of its implications for the past history of the east Bornean forests, it is essential to separate recurrent, if infrequent, phenomena that have been of evolutionary significance for these rain forests, from those that are products of recent anthropogenic land use practices. The following examination indicates that although severe droughts may recur, at this time no evidence suggests that forest fires of this scale have occurred before in lowland rain forest.

The historical incidence of drought in East Kalimantan can be examined indirectly, by comparing the monthly rainfall records accumulated at various rainfall stations in the province (Transmigration Area Development, 1983). For each of the 22 rainfall stations (excluding those in the wetter far north of the province), I calculated the percent deviation of each year's total rainfall from the station's long-term average, and then the mean among these deviations for each year. First, I examined whether the incidence of drought years since 1940 (when records from a sufficient number of stations became available) corresponds to known El Niño events in the western Pacific (Philander, 1983). Such an association is predicted from linkages between western Pacific ocean warmings that define El Niño, and the changes in the locations of low pressure areas and hence rainfall distributions which define the Southern Oscillation. In the 44 years for which rainfall data have been compiled since 1940, ten drought years have occurred (Fig. 6.4), of which nine are matched by one of the ten El Niño events recorded over this period (Philander, 1983). The probability of obtaining such a fit by chance is $< .001$ ($X^2 = 33.5$).

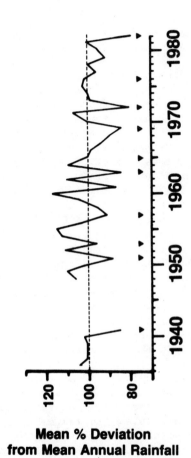

Fig. 6.4   Incidence of ENSO events since 1940 (solid triangles) compared with drought years in East Kalimantan (taken from Philander, 1983). N's for calculations of mean deviations range from 5–6 stations (1936–41, 1965–70), to an average of 16 rainfall stations (1972–82) (Data from TAD, 1983).

Figure 6.4 does not accurately represent the magnitude of various droughts, however. East Bornean droughts typically begin with unusually low rainfall during the drier months of June-October (Leighton, 1984; TAD, 1983), and in particularly bad years such as 1982-83, the drought persists into the early months of the following year. Consequently, total rainfall based on a calendar year may obscure a drought because of compensatory rainfall during other months. To evaluate the relative severity of the 1982-83 drought, we analyzed each year's June to May rainfall, for each of three widely-spaced stations in East Kalimantan (Fig. 6.1) that had been monitored for many years (Fig. 6.5). We calculated cumulative rainfall for all possible numbers of consecutive months and selected the minimum values to represent the worst drought measure for that length of time for that year. These distributions of drought intensities across years (Fig. 6.5) were then compared to the equivalent periods of least cumulative rainfall for each length of time during 1982-1983 (Fig. 6.1).

The hypothesis that the 1982-83 drought was particularly severe is supported. Minimum cumulative rainfall for various periods during 1982-83 (the arrows in Fig. 6.5) generally fall within or outside the extreme left tail of the distributions representing pre-1982-83 years. Although it is ill-advised to make conclusions from events that occur with such low frequency, the most reasonable inference to draw is that droughts as severe as that of 1982-83 occur on the order of once every fifty to several hundred years or so, and therefore are not without historical precedent.

The physiognomy of surviving drought-stricken trees in the burned primary forest at Mentoko provides direct evidence of whether a severe drought has occurred within the last 200 years or so. We assume that those trees showing severe crown dieback will for the remainder of their lives exhibit the characteristically asymmetrical, small bushy crowns they now have, as concluded by observers of Indian drought-stricken forest (Seth et al., 1960; Tang and Chong, 1979). We found no trees with these truncated crowns during our 1977-79 studies in this forest, except for the rare individual that had lost a portion of its crown when another tree fell. Since many if not most of the large canopy and emergent trees must be hundreds of years old (Meyer, 1974; Whitmore, 1984), the absence of such drought-induced truncated crowns prior to the 1982-83 drought argues that the current stand of canopy trees in the forest have not suffered from similarly intense drought stress. This implies that the drought of 1877 was not as severe as the 1982-83 one, because it left no record in the physiognomy of the numerous canopy trees alive in 1977-79 which would have survived the 1877 drought. It is possible that discrepancies between records of rainfall and tree physiognomy is due to factors contributing to drought stress in 1982-83 other than low precipitation, such as drying winds, low humidity and cloud cover, and high temperature, which need not be precisely correlated with low rainfall.

In light of these considerations of rainfall and tree physiognomy, it seems most prudent to adopt the tentative conclusion that components of moisture stress combine to result in massive dieoff of canopy trees from drought in East Borneo on a scale of once every several hundreds of years. Considering the longevities of canopy hardwood trees in these forests, such droughts constitute a potent selective force influencing plant distributions and relative abundances. For example, the predominance of Eusideroxylon zwageri and Koodersiodendron pinnatum in east Borneo (Meijer, 1974) may result from their drought-resistance.

Recent land use practices have greatly increased the liklihood that large-scale fires will follow a prolonged drought. The extent of slash-and-burn agriculture along the rivers and roads of the province has steadily increased during the last few decades. Thus, more agricultural fires could escape into the bordering secondary forests. Analysis of a series of thermal images from the NOAA-7 satellite has revealed that from a few fires at the end of February, the number grew to more than 100 by mid-April (Malingreau et al., 1985). Many fires persisted until mid-May, when only low-temperature, smoldering fires are seen, restricted to peat swamp forests (Malingreau, et al., 1985, pers. comm.). Furthermore, widespread logging during the mid-1960's through the late 1970's created large areas of selectively-logged primary forest situated between agricultural areas and primary forest (Gillis, 1984). Because fires readily burned selectively-logged forest, each area of primary forest was assaulted not only by broad fronts of fire, but in addition, by fires originating from different sources. (Leighton, 1984; Lennertz and Panzer, 1983; Wirawan, 1983, 1984).

Although the Kutai National Park was invaded by independent fires from several directions, most of these fires died soon after they entered primary forest from

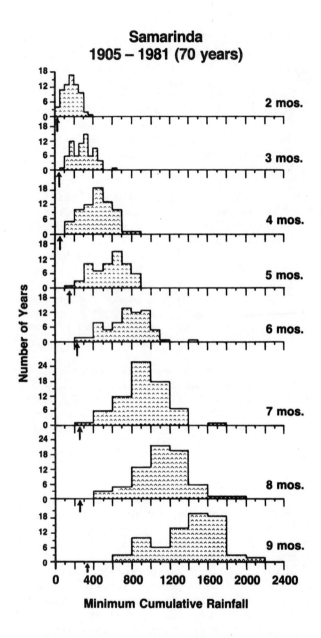

Samarinda
1905 – 1981 (70 years)

Fig. 6.5   Distributions of worst drought conditions (minimum cumulative rainfall, for various periods of consecutive months over all available years (see text) at three stations in East Kalimantan.   Numbers of years (in parentheses) at top are N's.

(continued)

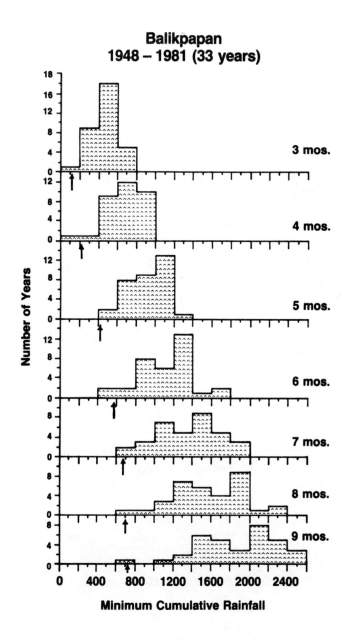

Balikpapan
1948 – 1981 (33 years)

Number of Years

Minimum Cumulative Rainfall

Arrows indicate minimum cumulative rainfall for each period during the July 1982 to May 1983 drought. To analyze a sufficient number of years, the Samarinda drought profile was a composite drawn from the nearby stations of Samarinda,

(continued)

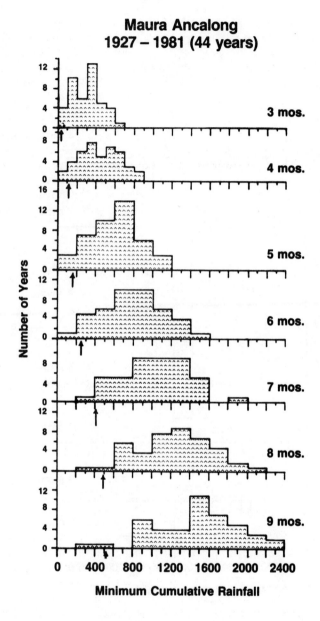

**Maura Ancalong
1927 – 1981 (44 years)**

Number of Years

3 mos.

4 mos.

5 mos.

6 mos.

7 mos.

8 mos.

9 mos.

Minimum Cumulative Rainfall

Sanga Sanga, and Sungei Kujang. Only one station's rainfall
was used for any given year. The Mentoko drought was
compared with the long-term data from Maura Ancalong, a
nearby station of similar seasonal profile (Data from TAD,
1983).

selectively logged forest. Much of the primary forest within Kutai did not burn, the drier Mentoko site being an exception. This suggests that under similar drought conditions in the past, fires were unlikely either to start or be sustained in primary lowland forest. Swamp forest may best resist burning in mild drought, but paradoxically, it is more likely that fires in peat swamp forests are self-sustaining, since the dried peat soil itself is combustible once it dries out. The fact that the smaller forest fires reported from peninsular Malaysia, Sumatra, and Sarawak all occurred in peat swamp forests suggests that fire may realistically have been a significant factor on an evolutionary time scale for this vegetation.

IMPLICATIONS OF THE 1982-83 KALIMANTAN FIRE AND DROUGHT

It is clear that tropical rain forests are dynamic systems subject to disturbances both small (e.g. tree falls, pest outbreaks) and large (e.g. hurricanes, landslides) (Connell, 1978), and that the frequency, severity, and extent of such disturbances influence how rich an assemblage of species can persist in the community (Connell, 1978; Hubbell, 1979, 1983). However, it is surprising and alarming that evergreen rain forests might occasionally suffer destruction from natural causes on such an immense spatial scale. The devastated area in East Kalimantan tragically includes the largest area of Southeast Asia currently covered in lowland evergreen rain forest, as well as the region's largest rain forest preserve, the Kutai National Park, in addition to being one of the world's richest rain forests (Whitmore, 1984). There was certainly no basis---either from historical records or from contrasting this rain forest with similar ones in the world---for predicting that such a calamity could possibly occur. The implication then is that severe drought, possibly in conjunction with forest fires, may be part of the history of other rain forests. This is in fact the case in at least one region of Amazon rainforest (Sanford et al., 1985). We should search for indirect evidence of this by examining soil cores or wood cores of old trees for charcoal layers (Sanford et. al., 1985). Severe disturbances occurring infrequently relative to the historical record (and to the length of ecological studies) may be overlooked as significant ecological factors or evolutionary selective pressures even though they have importantly influenced organismic characteristics and community composition.

The implications for conservation and land-use policy from the scale of this catastrophe are disturbing. Most large nature reserves are 100-1000 km^2 in size (Soulé, 1980; Terborgh and Winter, 1980), yet the area damaged by drought and fire in East Kalimantan covers 37,000 km^2. Susceptible populations within a nature reserve of typical size may be driven to extinction if a refuge is unavailable until the forest recovers sufficiently for populations to become reestablished.

Of immediate concern for East Kalimantan is the heightened risk that these damaged forests will be irrecoverably replaced by degraded vegetation, such as Imperata grassland, if another fire occurs in the next several years. These forests now contain large amounts of dead wood to fuel a hot fire and in most cases have understories dominated by softwood pioneer trees susceptible to burning (Kartawinata, 1980). The mortality among canopy trees has created a discontinuous canopy, making the forest prone to rapid dessication, so that even the drought-affected but unburned primary forests are now more susceptible to fire. It is therefore likely that over the short-term, conditions for even more destructive fires could be created by a drought of much less severity than that accompanying the 1982-83 ENSO event.

ACKNOWLEDGMENTS

ML thanks the World Wildlife Fund-U.S. for providing funds on short notice, the WWF-International for sponsoring the 1983 study, and the Ford Foundation for funding the 1984 research visit. The 1977-1979 research was sponsored by the National Science Foundation. NW thanks IUCN/WWF for funding and sponsorship. Prof. P. S. Ashton aided ML in several facets of the project. We thank the Indonesian government, the staffs of the Bogor and Samarinda PPA offices, Prof. Dr. Rubini Atmawidjaja, Dr. R. Bower, and members of the Forestry Faculty of Mulawarman University in Samarinda for permission to conduct this research and for sharing information with us. ML thanks L. and R. Berenstain for their hospitality and observations. C. Darsono, R. Samalo, and P. Pust and their families also assisted ML. T. Cuebas drew the figures. We thank P. S. Ashton, L. Berenstain, C. Folt, J. Glyphis, J. Mitani, D. Peart, and R. Primack for their for comments on various drafts of this manuscript.

REFERENCES

Asia Week. 1984. Wound in the World. 10(28), 32-49.
Barber, T. T. and F. P. Chavez, 1983: Biological consequences of El Niño, Science 222, 1203-1210.
Beaman, R. S., J. H. Beaman, C. W. Marsh, and P. V. Woods. Mscpt. Drought and forest fires in Sabah in 1983. Submitted for publication.
Berenstain, L. 1985. Mscpt.
Canby, T. Y., 1984: El Niños ill wind, National Geographic 165, 144-184.
Cane, M. A., 1983: Oceanographic events during El Niño, Science 222, 1189-1195.
Connell, J. H., 1978: Diversity in tropical rain forests and coral reefs, Science 199, 1302-1310.
Dayton, P. K. and M. J. Tegner, 1984: Catastrophic storms, El Niño, and patch stability in a Southern California kelp community, Science 224, 283-285.
Gillis, M., 1984: Harvard Institute for International Development, Development Discussion paper 171.
Hubbell, S. P., 1979: Tree dispersion, abundance, and diversity in a tropical dry forest. Science 213, 1299-1309.
Hubbell, S. P. and R. B. Foster, 1983: Diversity of canopy trees in a neotropical forest and implications for conservation. pages 25-42, In: S. L. Sutton, T. C. Whitmore and A. C. Chadwick (eds.). Tropical rainforest: Ecology and Management.
Kartawinata, K., S. Riswan and H. Soedjito, 1980: Tropical Ecology and development, 47.
Kerr, R. A., 1983: Fading El Niño broadening scientists' view. Science 221, 940-941.
Leighton, M., 1982: Fruit resources and patterns of feeding and grouping among sympatrice Bornean hornbills (Bucerotidae). Ph.D. Thesis, University of California, Davis.
Leighton, M., 1984: WWF-US No. 293 Project Report.
Leighton, M. and D. R. Leighton, 1983: Vertebrate responses to fruiting seasonality within a Bornean rain forest. pages 181-196, In: S. L. Sutton, T. C. Whitmore and A. C. Chadwick (eds.). Tropical rainforest: Ecology and Management.
Lennertz, R. and K. F. Panzer, 1983: East Kalimantan Transmigration Area Development Project PN 76.2010.7. Report.
Malingreau, J. P., G. Stevens and L. Fellows. In press. The 1982-83 forest firest of Kalimantan and North

Borneo: Satellite observations for detection and monitoring. _Ambio_.

Malingreau, J. P., 1985: Pers. Comm.

Meijer, W., 1974: _Field guide to trees of west Malesia_. Univ. Kentucky Press.

Philander, S. G. H., 1983: El Niño southern oscillation phenomena. _Nature_ 302, 295-301.

Rodman, P. S., 1978: In: G. G. Montgomery (ed.) _The Ecology of arboreal frugivores_. Page 465. Smithsonian Inst. Press.

Sanford, R. L., J. Saldarriaga, K. E. Clark, C. Uhl, & R. Herrera. 1985. Amazon rain-forest fires. _Science_ 227, 53-55.

Schreiber, R. W. and E. A. Schreiber, 1984: Central Pacific seabirds and the El Niño Southern Oscillation: 1982 to 1983 perspectives. _Science_ 225, 713-716.

Seth, S. K., M. A. Waheed Kahn and J. S. P. Yadav, 1960: _Indian Forester_ 86, 645- .

Soulé, M. E., 1980: Thresholds for survival: Maintaining fitness and evolutionary potential. Pages 151-170. In: M. E. Soulé and B. A. Wilcox (eds.) _Conservation Biology_.

Tang, H. T. and P. F. Chong, 1979: "Sudden" mortality in a regenerated stand of _Shorea curtisii_ in Senaling Inus Forest Reserve, Negeri Sembilan. _Malaysian Forester_ 42, 240-248.

Terborgh, J. and B. Winter, 1980: Some causes of extinction. Pages 119-134. In: M. E. Soulé and B. A. Wilcox (eds.), _Conservation Biology_.

Transmigration Area Development, 1983: Rainfall summary analysis for East Kalimantan, Samarinda, Indonesia.

Whitmore, T. C., 1984: _Tropical rain forest of the far east_, Second edition. Oxford, Clarendon Press.

Wirawan, N., 1983: IUCN/FAO Project No 1687, Report.

Wirawan, N., 1984: W. W. F. Project No 1687, Report.

Ghillean T. Prance

Afterword

A growing concern about the destruction of the tropical
rain forests of the world links the papers in this volume.
One important aspect of deforestation is addressed by the
authors: its affects on the atmosphere. However, this
subject cannot be presented without putting it in context
and realizing the gravity and extent of the deforestation
situation.

In the introduction, we are reminded of the species
diversity of the tropical forest habitat which contains at
least 10% of the total biological species in the world, and
also of the complex interactions that bind all the
organisms together into a stable unit or ecosystem. The
result of this diversity and the constant evolutionary
battle between predator and prey is that the rain forests
are also one of the most important sources of plants with
economic potential.

A worldwide summary of the depletion of tropical
forests is provided by Norman Myers. Deforestation is
indeed a serious problem when we learn that 200,000 square
km of this biome are being depleted each year from a total
area of less than 10 million square km. Both the chapter
by Myers and the following one by Woodwell et al. point out
the value of remote sensing to determine accurately the
rates of deforestation and also the speculations and
guesses presented in much of the literature. Woodwell et
al. show how rapidly the forests can disappear once a
region is selected for colonization. The Rondônia case is a
cause for alarm because it is also an area of extreme
biological diversity and local endemism.

The chapter by McElroy and Wofsy describes recent
progress in stratospheric and tropospheric chemistry with
an emphasis on its link to the biosphere. It also provides
a summary of current ideas on the importance of trace gases

103

in connection with climatic stability, and the sources and links of major trace gases with an emphasis on the tropical contribution. The disruption of tropical forests are of concern because they account for a relatively large amount of the world's biomass and an even larger portion of global primary productivity. Deforestation will have a profound effect on the budgets of atmospheric N_2, CH_2, and CO.

Salati et al. discuss the effect of deforestation on rainfall. Since something between 54 and 46 % of Amazonian rainfall is the result of the transpiration of water vapor from the forest, deforestation could have a profound effect upon climate. The significant changes in hydrology of deforested areas are also discussed.

It is significant that both the papers by McElroy and Salati et al. conclude with plans and ideas for future research. In spite of all the work reported in this volume, there are still many questions about the exact effects of deforestation. If we cannot do the research to answer these questions quickly, it will be too late both to save the species and to avoid the climate and atmospheric consequences that will occur.

In the final chapter, Mark Leighton emphasizes the precarious nature of the tropical forests. As the result of just a few years of exceptionally dry climate, vast areas of the forests of Borneo caught fire. It has often been stated that tropical rain forests do not burn, and in their normal and pristine state this is true. The shifting cultivator is normally safe when he sets fire to the patch of forest which he has felled. The fire scorches the trees standing at the margin of his clearing, but does not spread into the forest. However, in Borneo the combination of selective logging with the resulting debris in the forest and an exceptionally dry climate cycle meant that a phenomenal amount of rain forest was destroyed by fires. One cannot help linking this with the facts in the preceding chapter by Salati and realize that we are dealing with a sensitive and precariously balanced ecosystem which could very easily be completely destroyed if the current rate of deforestation is allowed to continue.

Should this happen, humankind will be the loser. One of the greatest resources of our planet is being squandered away. The saddest aspect is that much of the destruction is for short term gain that will do nothing for the future of the starving people of the world. The forests themselves are the clue and the source of sustainable yield systems in the tropics. If they are destroyed to create abandoned cattle pastures such as those along the

TransAmazon highway we have gained nothing and lost much more than may be initially apparent. The effects of the destruction will have a considerable impact on the atmosphere and climate of our planet.